Systems Analysis

IN THE SAME SERIES

Systems Analysis

Made Simple Computerbooks

Lyn Antill and Trevor Wood-Harper

MADE SIMPLE COMPUTER BOOKS

HEINEMANN : London

© 1985, L. Antill and T. Wood-Harper

Typeset by Wilmaset, Birkenhead, Merseyside
Printed and bound in Great Britain
by Richard Clay Ltd., Bungay
for the publishers William Heinemann Ltd,
10 Upper Grosvenor Street, London W1X 9PA

This book is sold subject to the
condition that it shall not, by
way of trade, or otherwise, be lent,
re-sold, hired out, or otherwise
circulated without the publisher's
prior consent in any form of binding
or cover other than that in which it is
published and without a similar condition
including this condition being imposed
on the subsequent purchaser.

British Library Cataloguing in Publication Data

Antill, Lyn
  Systems analysis.
  1. System analysis
  I. Title   II. Wood-Harper, A. T.
  003   QA402
  ISBN 0–434–98401–9

LIVERPOOL INSTITUTE
OF HIGHER EDUCATION

THE BECK LIBRARY

Accession No. 96290

Class No. STACK
004.21
ANT

Catal.

Contents

Preface

The purpose of this book is to introduce readers to the essential elements of information systems analysis and design, and to teach them some of the basic technical skills required for the tasks involved. It is assumed that readers will have some general knowledge of computers and their applications, but no particular assumptions will be made about either the level or content of that knowledge. This means that the book could be used as a primer by the relative newcomer, or for a technical update by either a businessman or a programmer.

The authors would like to acknowledge Professor P. P. Checkland, University of Lancaster and Professor Enid Mumford, Manchester Business School for some material in Chapters 3 and 5 respectively.

Introduction

Systems Analysis is the job of designing an Information System. The analyst first has to look at the information requirements of an organization, and particularly at any problems associated with the collection, processing, or storage of that information. He or she then has to design a system to meet those requirements. This will probably, but not necessarily, be based on some sort of computer.

The analyst is thus serving as a mediator between the organization and the supplier of computers and computer programs. This requires the evaluation of what is desirable against what is technically feasible. In fact the systems analyst's job is more complex even than this. Within the organization there will be several different views of the requirements. For example the production manager and the sales manager may not only want different information but may actually have quite different views about what information is vital to the company's success. At another level, the managers who want the answers will have quite a different attitude to the clerks whose livelihood may be threatened by a new computer system.

There are, in fact, five aspects to the design of an information system:

clarifying the purposes of the organization, or some part of it,
the analysis of the information itself,
the design of a socio-technical system so that computer operations can be fitted into daily work,
the design of the man-machine interface to enable individuals to relate to the computer,
the creation of a computer system to support the chosen design.

In the past different systems analysts have taken a different approach to their job depending on whether they started life as computer programmers, users, management consultants, or specialists in industrial relations. But the discipline of systems analysis has grown up. A framework now exists to tackle all the aspects of an information system. This does not solve the problems, but it enables decision makers to see what the alternatives are and what are the implications of each.

Computer technology will change continuously in the foreseeable future, and new techniques will evolve to deal with the details of the analyst's job. However, the essentials of the analyst's role are now clear. A basic understanding of this role will enable the future analyst to adapt to technological advance.

Systems Analysis is a practical discipline. One must understand the principles and know how to carry them out in practice. This book sets out that basic role and uses the latest techniques to illustrate how it may be carried out. It concentrates on the first four aspects because the design of the computer system will depend largely on available technology, and will therefore change in a way that the other human activities will not.

The main examples in the book are from small systems, both because they give a good overview of the analysis without overwhelming the reader with the sheer volume of work, and because a novice systems analyst may well be required to tackle a small systems design on his own. Trainee analysts in large organizations are likely to be occupied with detailed tasks for some time before they progress high enough to see the whole picture again. The systems discussed are for a video shop, a professional institute, and an air-freight company.

1
Information Systems and the Organization

Before looking at the processes of systems analysis and design, it is necessary to understand something of the ways in which information is used by various types of commercial and administrative organizations. This is because we must know something of the problem before we can try to solve it.

1.1 TYPES OF INFORMATION PROCESSING ACTIVITY

While it is true that every human situation is in some way unique, it is nevertheless possible to recognize certain types of activity which have similar information processing requirements. Recognizing that the activities with which he is working are of a particular type is a useful start for the analyst, and enables him to structure his enquiries in a more efficent way. Saying, for example, that 'this is a medium-sized company engaged in retailing high value goods' does not tell you what computer system they require, but it does suggest some of the questions you should ask, e.g. 'what information do you require about these goods?' 'Since an individual transaction is of high value what security measures do you need to take to ensure each transaction is properly processed?' It may also suggest similar organizations where the same sorts of problems were tackled, e.g. selling Rolls-Royce cars has similarities to selling valuable antiques in that you cannot afford to have one sale recorded incorrectly. In a supermarket on the other hand you could expect that errors may occur perhaps on one item in a hundred, or possibly one in a thousand, and that minor inaccuracies have to be tolerated.

Here are some frequently encountered types of organizational activity, each with its own information processing requirements. There will be differences between two organizations of the same type, or between a large and small organization doing similar work. There will also be organizations which do not fit into any of the categories shown. One organization may be engaged in several of the activities.

1.1.1 Order Processing

Many types of company are involved in accepting and filling orders. One example would be a mail order company which collects orders through the post or over the telephone. These are checked to ensure suitable arrangements for payment. The order is then checked to see whether the goods are in stock. If so they are despatched to the customer. If not they may be ordered from the supplier or the customer's payment may be refunded depending on company policy. If payment does not come with the order then an invoice must be created. Either way the accounts must be kept.

Another order-processing organization is a factory parts-store. The storeman receives requisitions from around the factory. These do not have to be paid for but they do have to be properly authorized. If the parts are available they are supplied, if not arrangements must be put in hand to order them from suppliers, have them made up, or supply acceptable alternatives.

A pharmacy is another order-processing system with the prescription doing the job both of order and authorization.

Essentials of an Order-Processing System

1. Orders come in.
2. They are checked for payment/authorization/credit-worthiness.
3. If the ordered items are available they are supplied.
4. If the ordered items are not available they may be ordered or made up.
5. Where goods are supplied on credit, an invoice must be sent.
6. Records are kept of stock in hand or on order.
7. Accounts are kept of actual payments by customers, or for internal costings.

Other order-processing systems:

the stock room of a shop,
a warehouse supplying retailers,
a warehouse supplying a chain of supermarkets,
a company supplying made-to-measure items—curtains, furniture etc.,
a carpet shop,
a shop selling self-assembly furniture.

1.1.2 Booking Systems

These are rather like order entry systems in that requests come in from outside which the system must endeavour to meet. The difference is that there are no goods to be supplied. Instead there is a customer requiring a seat on a plane or at a theatrical performance, a patient needing an appointment with a dentist, or a client wanting to see his lawyer or bank manager. The major difference from the order-processing system is that what one is trying to book is inherently a scarce resource. This means that one has to check the availability of the seat or the appointment before accepting the booking. There is likely to be a discussion with the client about various alternatives available.

1.1.3 Administrative Systems

Administration is about keeping track of what is happening in an organization. Accounting is the most obvious administrative function. This involves keeping records of payments and receipts, including staff salaries and other overheads, as well as trading activities. It may also involve apportioning costs between different parts of an organization to check on the profitability of different activities. The Accounts function provides a service to the profit-making activities of the company (e.g. order entry) in recording payment and chasing up invoices. It fulfills a legal requirement because all companies are required to keep proper books of account which have to be certified by auditors and submitted to the Inland Revenue for tax purposes and to Customs and Excise for VAT. This is known as Financial Accounting.

Finally Accounts provide a service to senior management in showing up which areas of a company are most profitable. The same service is provided to a non-profit making organization (e.g. hospital, college, government department) by showing which activities are most costly.

Another administrative function is the keeping of inventories. How many desks, typewriters, filing cabinets does the company own? This may be thought similar to the order-processing system in, say, a warehouse. There certainly is some overlap, but there is a fundamental difference. A warehouse system is designed to cope with a high turnover of goods, an inventory system is just to keep track of items which happen to have been acquired or disposed of. The function of an inventory system is simply to report, so that capital assets may be accounted for and managerial decisions made about future acquisitions and dispositions to ensure the most profitable use of capital items.

A personnel system processes hire, fire, and wages decisions made throughout the company. It records personal details of employees, salary levels, tax and bank account details, hours worked, expenses claimed. It must comply with the Inland Revenue, auditing, and employment legislation requirements. It provides a staffing service to the whole organization, and can provide to management information concerning decisions about staffing levels, wage bargaining etc.

Essentials of an Administrative System

1. Activities are initiated elsewhere in the organization.
2. These activities are recorded.
3. The nature of the records is determined by regulating bodies.
4. The system provides a service to those parts of the organization actually providing goods and services.
5. The system provides information which can be used for management decision-making.

Other Administrative Systems

Licensing (cars, public houses, lotteries) except for the initial decision on whether or not to grant the licence.

High street banking (except for decisions on whether or not to grant overdrafts, mortgages etc.). This is quite different from Merchant Banking which has aspects of order-processing and real-time control.

1.1.4 The General Office

While many of the other information processing functions are performed in an office, interest has recently been taken in general office activities. Correspondence, reports, filing, minutes of meetings, diaries, etc. Message passing is of great importance where staff have to travel, e.g. salesmen. Minutes of meetings are important in public bodies, e.g. local government.

Office systems have to be flexible and responsive to the needs of the situation. Varying degrees of standardization are possible in office systems, e.g. standard letters or paragraphs, report formats, mailing lists, memo forms.

1.1.5 Decision Support Systems

When a bank manager 'decides' whether to authorize an overdraft what he is actually doing is using his discretion in implementing a *rule*. The decisions about overdrafts are made, at least in a general way, by Head Office. A clearer example of this was the decision of the High Street banks around 1980 to go into the mortgage market. After the decision had been made, guidelines were issued to branch managers. An individual branch manager could not have made this decision himself. Similarly, a *salesman* authorizing a hire purchase agreement is implementing a company policy, whereas the business *owner* decides whether to make a credit arrangement.

It is important to make this distinction because there is a fundamental difference between the information requirements in the two cases. Even though the branch manager may have to use his judgement in an individual case, he has a set of rules to guide him, e.g. Does the customer have a good repayment record? Will he have sufficient funds for repayment? What risk is involved? When the entrepreneur decides whether to invest in fashions or new technology, he has few, if any, guidelines to work to. Moreover he is perfectly justified in ignoring the guidelines and following his 'instinct'. This is often where the most spectacular advances are made.

The most obvious thing about decision support systems is that they are only semi-structured, because as soon as guidelines have been produced the function gets absorbed into the administrative system. The information requirements are determined by what the decision maker wants to know. Analysing existing company data provides some answers. Others may only be obtained by research, e.g. the activities of competitors. The greatest difficulty is in trying to analyse what information it may be worth keeping about current activities for the sake of future decision making.

For example the government were under pressure to do something about supposed inequalities in the education of ethnic minorities and girls. Existing data was analysed to provide evidence of the relative failure of girls in mathematics and science (and incidentally of boys in languages). Evidence of a political nature was collected on the views of different groups. Information on the relative performances of different ethnic groups was not available in existing records as it had been for the girls because no records were kept of ethnic origins. This was partly because no-one had thought it necessary to collect such data, and some people thought it was not appropriate. It was also because ethnic origin is not a matter of simple calssification, it is also a matter of the individual's perception of himself. A third-generation black immigrant may feel more English than a white person whose family had lived for three generations in Africa. The DES required schools and colleges to provide information on the education of ethnic minorities but it was not so clear how this should be done or whether such records should be kept routinely in the future.

Many decisions were made by the DES and by individual authorities and colleges before all the evidence had been accumulated. They were based on political judgement or examples of glaring inequality. Information was used to support decisions. The decisions were not based on a rigorous analysis of existing information.

Essentials of a Decision Support System

1. Decision situations are only semi-structured and it is difficult to predict what information will be required.
2. The information requirements depend on what the decision maker wants to know at the time.
3. Some of the required information can be obtained by analysis of existing records.
4. Some can only be obtained by research.
5. Some information might have been included in existing records if people had known that it was going to be useful.
6. There is often a time limit by which a decision of some sort must be made.

1.2 COMPUTER BASED INFORMATION SYSTEMS

Just as no two organizations are identical, so neither are any two computer based information systems. Even when two companies use identical machines with the same basic software, the database, applications programs, and associated clerical and operational procedures will differ. These are all essential parts of what is often called the 'computer system'. This is not a very good term because it concentrates on the machine rather than the human activities. 'Computer based information system' is more accurate but it is rather long. The abbreviation CBIS will be used. 'Computer system' will be used for the hardware and basic software.

There are several different general types of computer system which can be used to support different types of CBIS. These are not absolute categories but nevertheless represent useful distinctions. More specific technical differences between types of computer system may be of concern to the programmer but not the user or the analyst.

1.2.1 Mainframe, Mini, Micro

This refers to the type of computer at the centre of the system. The most obvious distinction between these is size—large, medium and small respectively. Large also means fast at least in terms of total throughput, and also powerful in the size of calculations that can be performed or the number of different tasks that can be performed simultaneously. Another obvious difference is in the environment in which the computer works. A mainframe usually requires an air-conditioned room with a specialized power supply. It requires a reinforced floor raised to permit all the cabling to run underneath. It also requires specialized operators in addition to data entry clerks at the VDU terminals. There may be a hundred or more terminals used at the same time. They may all be feeding data into one program, or each doing its own job. Apart from sheer size the main feature of the mainframe is its sophistication and complexity.

A mini will usually work in a normal office environment with only minor adjustments to the electricity supply. A specialized operator may not be required, unless the system is frequently switched between different jobs. Often all the necessary operations can be performed by users after an appropriate training course. VDU operators still have to be shown how to use each application. There may be dozens of terminals and several different programs running at once. Arguably the greatest strength of the mini is its use with programs designed to perform certain types of work (e.g. order processing) but which can be adapted to meet particular requirements, such as simple ways of defining what details should go on each order, what each invoice should look like etc.).

The micro hardly needs describing because it is so well known. At the bottom end it is little more than an adjunct to the domestic television set. At the top end it is in some way more sophisticated than a mini. Micros lack the complexity of the larger computers and so they are easier to program. There are also many ready-made programs ('packages') available. These offer new facilities like spreadsheets, word processing and business graphics as well as simplified versions of minicomputer software such as order-processing a database.

1.2.2 Office Automation Systems

A word processor is essentially the same as a micro, but with a text manipulation program built in. Microprocessors are built into switchboards to enable intelligent re-routing of calls. Word processing facilities can be built into telex machines to enable cheap, accurate, and efficient transmission of messages worldwide.

Microcomputers can provide word processing and local computing power. These can be linked to each other or to larger computers to give the possibility of transmitting raw data or processed information to other users.

Electronic typewriters enable faster and neater production of documents. Memory devices and optical character readers enable typewriters to be interfaced to word processors.

Electronic mail is a new message passing system enabling word processed messages to be left in a specified area of a computer store known as a mailbox. This message can be collected by the authorized user from his computer terminal. This terminal may be portable and connected via an ordinary phone.

Other technologies are being integrated into office automation, particularly reproduction and document transmission. Printing can be done straight from a word processor disk or from camera-ready copy produced by a word processor. Facsimile transmission is a method of transmitting an image of a document by telephone line and reproducing it. This is expensive at the moment but when the technology is fully developed it may cause some rethinking of the communications process.

1.2.3 Telecommunications

Communication between computers is playing an increasingly important role. This is achieved in two main ways. A Local Area Network (LAN) connects computers within a single site. A wide area network uses telephone lines to transmit data to other sites. This may be a leased line or it may be a public line possibly using a British Telecom Service such as Packet Switching (PSS). The computers may be independent of each other and merely exchanging information, or one computer may act to control the data distribution. The computers may be connected in rings, stars or other more complex arrangements.

Private Automatic Branch Exchanges (PABXs) which have been developed from more conventional telephone switchboards are now becoming computerized message passing systems trans-ferring calls from one extension to another. PABXs can carry voice and data.

Broad bandwidth cable can carry voice, data, TV pictures and also signals from things like burglar alarms. Thus it is possible to have a whole variety of previously distinct information types carried along the same cable. The implications of this are only just beginning to be explored.

Satellite communications are being rapidly extended so that greater volumes of both voice and data can be carried. Inter-national standards are being developed to facilitate the design and implementation of communication systems. Standards for communication between one make of computer and another, and between one set of programs and another are much more contentious.

1.3 SUMMARY

There is a variety of information processing activities that can take place in an organization, and there also is a variety of different types of Information Technology that can be used to support these activities.

Information Processing can be roughly categorized as follows:

order (or transaction) processing,
booking,
administration,
general office work,
decision support,
telecommunication.

Computer systems can be divided roughly into mainframe, mini, and micro according to their size. Additionally a computer may be 'stand-alone' or linked to other computers. These links may be over a telephone line. Users may also have access to distant computers via a terminal and a line. Microcomputer technology is being incorporated into other office equipment such as word processors, switchboards and telex machines. It is increasingly easy for information to be transmitted world-wide via telephone lines and satellites.

2
The Role of the Systems Analyst

Before going on to look at the techniques of systems we need to consider the role which the systems analyst plays in an organization. Let us start with a very general definition:

'The systems analyst works with the user to specify the information requirements of an organization and with suppliers of hardware and software to implement a system to meet those requirements'.

This conveys the intermediary aspect of the role. The title Systems Analysis and Design is often used to convey the creative aspect of the role. There is a sense in which the analyst is like an architect producing designs to the client's specification, or for his approval, which can then be turned into an actual construction by the builder.

There are many ways in which this role can be carried out and we will consider the major ones. They are influenced by such things as the size of the organization, the particular job in hand, the historical development of computing in the company, availability of suitable expertise and the preferred style of senior management.

2.1 THE 'TRADITIONAL' DATA PROCESSING DEPARTMENT

For some years now DP has been regarded as a 'service' function with the same sort of relationship to the rest of the company as Personnel, Accounts, R & D. The justification for this is obvious where there is a large central computer. This is an expensive resource, needing a variety of experts to operate it, and providing a service to other departments in the organization.

Answerable to the DP manager there will be operators, programmers, and systems analysts. There may also be data preparation staff who do keyboard entry of data, although much of this is now done directly by users at their own terminals. Between the three functions of operations, programming, and analysis there are various ways of subdividing the tasks to be done. Within systems analysis the tasks are:

user liaison,
information strategy,
data and function analysis,
database design,
program design and specification,
hardware selection,
design of clerical procedures,
testing,
'systems' implementation.

It is unlikely that all these tasks would be handled by one person except in the smallest organizations. For example user liaison and the design of clerical procedures can be carried out by people from user departments, and strategic decisions may be made by the Board. In some companies analysts will do much of the database design or program design. They will work alongside programming staff.

The division of responsibilities will be reflected in the organization of the DP department. It may be split along functional lines. The analysts would then create a design in a specified format which would be handed over to the programmers. The other alternative is to assign both analysts and programmers to project teams. The team would then be responsible for all aspects of the implementation of a particular DP project.

The functional split maintains the highest level of technical expertise and facilitates better standardization across all DP projects, but it is prone to systematic misunderstandings at the point where the analyst hands over to the programmer.

The team approach aids the integrity of a particular project but is prone to differences between one project and another both in technical details and in the manner in which the user interacts with the system.

It is necessary for a DP department to select the most suitable organization for its own circumstances, (which includes the question of career progression for DP staff) and then to guard against the inherent dangers of that organization. However, the very existence of a DP department provides problems of its own which have become increasingly obvious with the growth of on-line and distributed computer systems. This is the problem of the relationship between DP and the departments whose operation depends increasingly on the computer. Various arrangements can be made to alleviate these problems such as committees with user and DP representatives, secondments of users to analysis or project teams, or secondments of analysts to user departments to help in implementation or operation of new systems.

2.2 USER COMPUTING

With the advent of microcomputers and of minicomputers with ready-made software has come a split between users and DP departments. Operators and programming staff are not required if all you are going to do is tailor accounting, data management, or other packages to meet the needs of users. There is still a need for analysis in this situation although this is often undertaken by the users thenselves. This is because many analysts are seen as being too concerned with the technicalities, and not being sensitive enough to the actual needs of users.

Judging by the rapid increase in work on packages, software engineering, expert systems and silicon chips, a computer system will be more and more able to take care of its own technicalities and to be much easier to use. This means that analysis will be less technology dependent and more concerned with the needs of users.

The DP department will not cease to exist, but a large number of new analysts will be needed to work directly with users. They will be concerned with Information Strategy and user-liaison, data and function analysis, and the selection of equipment and software packages. They will also deal with tailoring packages to the specific needs of users, training in their use, and will advise on the organization of clerical work around the computer.

2.3 THE CONSULTANT ANALYST

We have so far shown two ways in which the analyst may perform his or her role within an organization. However there are situations in which organizations may need to call in an analyst from outside.

The first of these is the contract analyst who must quickly fit into the organization and get on with a specified task. This is usually done to cover a temporary gap in DP staff and will not be considered further, although quite a lot of analysts earn their living this way.

When an established DP department is considering moving into a new area they may want to call on high level specialist advice. Senior management may want an impartial discussion of company strategy with regard to information processing. A feasibility study may be needed of a potential development. A consultant may even have to act as an unofficial arbitrator between conflicting interests within an organization, who have different views of the way in which Information Processing should be handled.

To act in this role the analyst must have acquired considerable experience of different organizations and computer installations. The analysis will not be detailed but it will require an overview of all aspects of the system; human and organizational as well as technical. To aspire to high-level consultancy work the analyst must have experience of all facets of the profession, and also have learned when and how to call on technical or business specialists for advice and information.

As the price of computing falls, so does the bottom level of consultancy. A small business or an individual department within a larger organization may well call in a consultant for a few days or weeks to help them select a micromcomputer with appropriate software and to set up a system based on it. Again this calls on all aspects of analysis but in a simple and straightforward way. Many niggling difficulties may be encountered, but few theoretical ones. Also missing are the inherent difficulties of co-ordinating a large team.

The microsystems consultant has only emerged in the eighties. Some may be members of a professional organization such as the British Computer Society or the Institute of Data Processing Management. However, these are not statutory bodies and cannot enforce codes of practice. Many so-called consultants have no professional or academic qualifications, although they may have practical experience and/or detailed knowledge of particular micros and packages. In the foreseeable future some licensing of consultants is likely, although it is not clear what form this might take.

The consultant must set up a clear contractual relationship with the client so that the client knows what he is paying for and the consultant knows what he is supposed to do.

2.4 THE SOFTWARE HOUSE

Companies who do not have a DP department, or whose DP department lacks time or particular expertise, may call in a software house to undertake a project or to inject specialist skills into a project. This activity can overlap that of the consultant. The software house may be linked to the equipment supplier.

The software house must also have a well specified contract with the client. This is particularly so because the software house is producing something for the client—either a systems design or a set of programs. The difficulty is in testing whether the design meets the client's agreed requirements. Various measures can be adopted to agree a design stage by stage and to specify acceptance criteria for the completed project.

Once the contract is agreed, the work proceeds in much the same way as for an in-house project. There may be additional difficulties of communication where the software house is on a separate site from the user, although many DP departments are physically separate from users within the same company, so the same problems can exist.

2.5 THE EXPERT WITNESS

New technology laws and agreements in Scandinavia call for and sometimes fund technical experts to advise workers on the implications of technical decisions about proposed computer systems. This role for the analyst is not common in the UK but the growing interest of clerical unions in new technology agreements indicate that more analysts may be called on to work in this role.

The intention is to help the people who will be using the technology to comprehend the way in which it will affect their working lives.

2.6 SUMMARY

The role of the analyst is to help the user clarify his information processing requirements and to choose the most suitable general systems design to meet these requirements. The analyst must perform the detailed analysis and work with programmers and others to help to implement a working system.

This role, or some parts of it, may be carried out from different positions within the organization, or from outside it:

within a data processing department,
inside a user department,
as a consultant to the organization,
within a software house retained by the organization,
as an expert witness giving advice to, say, a trade union.

3
The Human Activity System

The purpose of creating an information system is that it should meet a need. In other words there is a problem which will hopefully be alleviated by better information processing.

The problem exists within a human activity system (HAS). That is a group of people working together. This system may be a company, a government department, a sports club, a single office within some larger organization. It may be one person, a sole trader wanting to keep business records, or a teacher recording student progress. The analyst must understand the HAS in order to study the information flows involved in it.

Thus the first job of the analyst is to help the people in the organization define the situation and analyse what the problem is before they can set about solving it. The analysis of the human activity system consists of clarifying what group of people and activities are being discussed, defining the key elements of the system, pointing out any conflicts of interest, and coming to an agreed definition of the problem that is to be tackled.

3.1 THE RICH PICTURE

The first problem the analyst encounters is that he or she may have no knowledge of the HAS. Funnily enough this is often less of a problem than the fact that the people involved will each have a different perception of what the HAS is all about.

The analyst coming fresh into a situation has a chance of being reasonably objective, whereas the people already in it will have developed their own ideas which will often be unstated. Not only will everybody have a bias towards those tasks in which they are involved, they will also have different assumptions based on background, education, political persuasion, temperament, etc. Unless these differences are clarified they can lead to some awkward misunderstandings. Not many people will have stood back and looked at their organization in this sort of way before, and may be surprised at what they see in the picture.

The analyst's relative ignorance can be turned to great advantage as he or she works to draw up a picture of the activities, conflicts, and priorities which can be agreed by all concerned. The pictures are simplistic and may be humorous, but they form a solid basis for further analysis because of the way they set the scene and ensure that everyone is talking about the same thing. Working on them also helps to establish a proper relationship between problem owner and problem solver as they come to understand how they can help each other tackle the problem.

The examples shown here illustrate more clearly than words what is required in the picture of an HAS. The crossed swords indicate conflicts of interest. For example in the Video Shop the accountant would like detailed records of every transaction, whereas the shop assistants want minimum book-keeping hassle so they can concentrate on faster turnover. This was fast turning into a battleground. The thought bubbles show the main concerns of the people involved. You can also see the way in which the manager and the competition were eyeing each other closely.

It may take several attempts and discussions with representatives of the different actors before the picture is agreed. However this is time well spent because all further analysis work can be more surely directed towards the agreed problem.

3.2 THE ROOT DEFINITION

Different people who have a stake in the system will have different opinions about it. If you ask each of them questions like 'what is the main purpose of your company?', you will get different answers, such as:

'to make a profit'
'to keep people employed'
'to provide a service for the customer'.

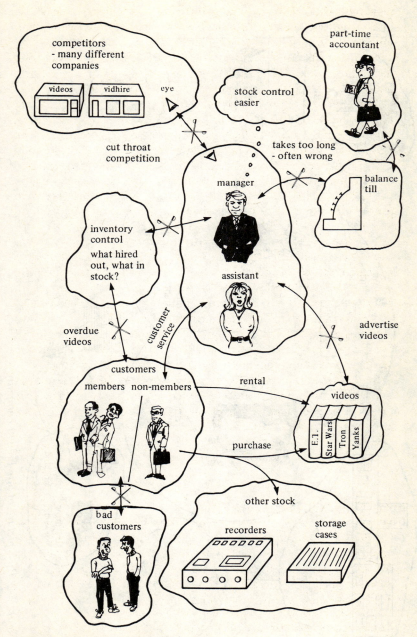

Figure 3-1 Video Shop Rich picture

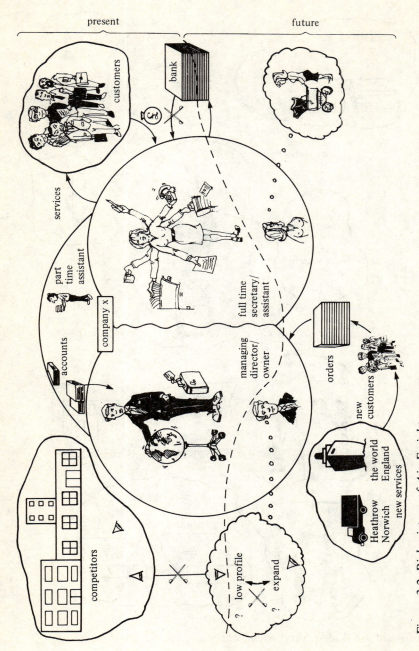

Figure 3-2 Rich picture of Air Freight company

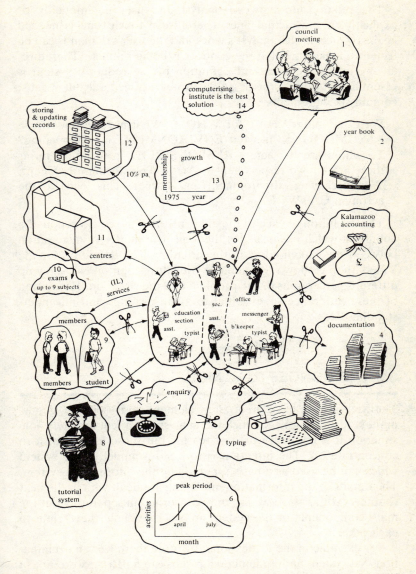

Figure 3-3 Part of a Rich picture of a Professional Institute

These are all valid statements of aims, but they have conflicting implications for the organization. Also they are much too vague to help the analyst produce a system which will help the organization in furthering its aims.

The relationship between analyst and user is, implicitly or explicitly, a contractual one. Therefore a touchstone is needed against which all the analyst's work and the user's demands can be tested to make sure that the contract is being fulfilled. Therefore, at an early stage a careful definition of the required system is essential. Of course this is going to be very general but it must contain six ingredients:

WHO is doing WHAT for WHOM. What are the underlying ASSUMPTIONS? in what ENVIRONMENT is it happening? To whom are the actors RESPONSIBLE? In technical terms these are known respectively as Actor, Transformation, Client, Weltanschawng, Environment and Owner. This leads to the acronym CATWOE.

The definition of each of the elements, and the construction of a definition which encapsulates them all, is a matter of negotiation between the stakeholders in the situation. The definition for the Video Shop is shown. Arguments over the definition will show up different perceptions of the problem and the situation. Resolution of these differences is essential to avoid disputes and misunderstanding later on.

3.3 THE CONCEPTUAL MODEL

The Rich picture has a great many elements in it—people, goods, money, ideas, conflicts. Once the analyst has a feel for the whole situation it is necessary to start teasing out those aspects of it to concentrate on. The intention is to build a model of the system which can be used as a basis for studying the information system. The details of the information modelling are dealt with in the next chapter. At this stage we are only concerned in separating out the main components of the system and to show how they relate to each other.

For example in the Professional Institute there was an examinations subsystem and a membership subsystem. These were not in isolation because members were involved in various capacities in

Root definition

C customers
  - Video shop staff
A actors
  - Video shop staff
T transformation
  - To provide cost effective stock control,
    customer record control, and easier
    accounting facilities
W weltanschawng
  - Low cost, efficient
O owner
  - Video shop owner
E environment
  - The area where the shop is located

A system to provide low cost and efficient stock control,
customer record control and assistance to   accounting facilities.

Figure 3-4 Video Shop

examination. Nevertheless they could be distinguished and the
relationship between them studied. There was also an accounting
subsystem. One could have done some accounting within the
examinations system and some within the membership system, but
cheques, invoices, etc. from both systems were passed over to the
accounting system.

We also have to remove people, although not their roles, from
the picture, because we do not want to create a system around
particular personalities. Later on we will look at how to fit a
computerized information system into the lives of the people who
will be using it. This may involve staff changes or changes in
individual jobs so we cannot build particular people into our
model. All we are concerned with now are the tasks those people
are performing.

In many organizations some of the staff may be wearing more
than one hat, as it were. Thus someone could be serving a
customer one minute and checking stock another. This must not
confuse the fact that serving and stock checking are different
functions calling for different information processing activities.
This is another reason for separating tasks which must be done,
from the people who happen to be doing them.

Figure 3.5 shows a conceptual model of the Video Shop. Note
how it has been derived from the corresponding Rich picture in

fig. 3.1. The boundaries show the various subsystems within the overall organizational system. As with the Root Definition, this conceptual model must be agreed.

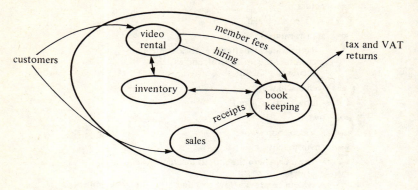

Figure 3-5 Conceptual model for a Video Shop

3.4 SELECTING PRIORITIES FOR ANALYSIS

Even the smallest organization will turn out to be composed of several subsystems and it is quite likely that you will not be able to tackle all of them at once. It will probably be necessary either to select out particular areas for attention or to decide on a sensible plan for tackling the job one area at a time.

The first person who has a say in this is the problem owner. The analyst may well have been retained specifically to tackle 'the Accounts' or 'stock control' or 'vehicle scheduling'. The only thing that is left for the analyst to do is to define the boundary between the activities to be included in the study and those still outside. The exact position of the boundary must be a matter for discussion between the analyst and the user when more is known about the details (see Chapter 4). Only if the analyst is convinced that there are real difficulties with the user's choice of system to be tackled should the matter be pursued further.

Often, however, the analyst will be asked to advise on areas of the company's operations where a new information system could produce the greatest benefits. The area that the analyst selects may be the only one to be tackled, or it may be the first part of a phased study of information processing throughout the organization.

A variety of factors have been shown over the years to indicate the suitability of an area of activity for computerization:

many repetitions of well-defined tasks,
large volumes of well-defined data to be stored and retrieved,
complex, but well-defined, calculations to be performed on data sets of any size,
extensive collating, searching or sorting of data,
people in different locations needing access to the same data.

Recently new candidate activities for computerization have occurred:

where a manager needs to access selected data and then try out 'what if' calculations on it to assist in decision making (e.g. what if prices went up by 5 per cent, what would our profits look like?),
where standard text had to be used and incorporated into documents (e.g. legal contracts, insurance policies),
where a lot of information needs to be circulated quickly to people who will not normally be at the other end of a telephone but could get to one to dial in to an electronic mailbox,
complex decision-making situations where expert rules can be deduced and built into a computer system to guide the inexpert.

The fact that an activity *could* be successfully computerized does not mean that it *should* be. The assumption is usually that the analyst has been called in because there is a problem. The roots of the problem must be exposed to see whether improved information processing would improve matters. The expected improvement would have to be enough to cover the cost in time, money, and hassle of implementing the new system.

Sometimes a problem cannot be solved by computer power, particularly where important data was impossible to obtain in the first place. The analyst may be able to suggest areas of activity which would seem far more likely to yield improvements in performance than those he was actually called in to investigate.

As far as the organization is concerned it is necessary to tease out those areas of activity which are critical to its success or which are using up the most resources. These are the ones which are most likely to show a return on any investment in computing. This is simply because a 10 per cent improvement on an area representing 50 per cent of the costs is worth the same as a 25 per cent improvement on something only using up 20 per cent of the costs.

Thus the analyst might be advised to put some ££'s on the Rich picture and to ask questions like the following:

Is there a lot of money tied up in stock?
Could the cash flow be speeded up to reduce total indebtedness?
Is there a lot of staff doing routine clerical jobs and typing?
Can managers get at the information they need for planning and decision making?
Are significant numbers of customers being lost through goods or services not being available when required?
What are the most expensive parts of the operation?

3.5 SUMMARY

Before the analyst can look in detail at information processing he or she must have a clear idea of the organization in which that activity is taking place. This can be done by developing a Rich picture of the organization. This displays the main components, activities, personalities and their relationships. It also highlights conflicts and areas of concern.

A root definition of the required system must be agreed with the users. This covers *who* is doing *what* for *whom*, and *to whom* they are responsible, in what *environment* this is happening, and what *assumptions* underpin that activity.

The root definition can be stripped of things and people to form a conceptual model of what is going on, and from this can be selected priorities for activities that might be worth computerizing.

4
Information Modelling

Once the analyst and the problem owners have agreed on the overall picture of the situation and on the root definition of the system to be designed then the process of abstraction begins. The analyst must work with the users to determine the WHAT of the new system:

WHAT records are to be kept
WHAT processing is to be done
WHAT events trigger those processes.

No attempt should be made at this stage to consider HOW this should be done. It is very tempting to think of particular databases, application packages, and programming languages which appear suitable for the application in hand. However, even a suggested commitment at this stage would pre-judge the issue. Also, the whole information model can be created and manipulated quite easily on paper. Indeed it is much easier to change a paper model than a computerized one, and there is every chance that several rounds of changes will be necessary before everyone is satisfied.

The final part of the modelling process is to decide which of the organization's activities are actually to be included in the model.

4.1 ENTITIES, FUNCTIONS, AND EVENTS

These are the cornerstones of the model and so it is important to be sure what they are. Probably the least familiar of these terms is ENTITY.

An ENTITY is a thing that you want to keep records about.

This might seem rather vague and arbitrary. In fact there is a great deal of flexibility in the selection of entities in any particular model. You can keep records about:

people, (e.g. customers, staff, clients, patients),
transactions (e.g. purchases, loans, shipments),
things (e.g. parts, stock, books),
events (e.g. births, marriages, and deaths),
accounts,
anything else that is important to the organization.

Entities have ATTRIBUTES associated with them. Thus a customer may have a name, an address, etc.

FUNCTIONS are all the things that have to be done. They might involve collecting, using or amending information. Examples of functions are:

produce invoice,
calculate profits,
admit patient,
record student marks,
control stock.

Functions are naturally hierarchical, i.e. major functions such as 'control stock' consist of functions such as 'record deliveries', 'record shipments', 'estimate shrinkage', 'calculate optimum stock levels' etc.

EVENTS are things which happen either inside the information system, or in the outside world, which trigger functions, e.g. a customer wanting to buy something is an event which triggers the function 'process sales'. This may in turn trigger off other functions—'check credit rating' or 're-order goods'.

Some systems are much more 'event driven' than others. Reservation systems and process control (oil refineries, robots) are examples of event driven or real time systems.

4.2 FUNCTIONAL DECOMPOSITION

For business and administrative systems the creation of the information model usually starts with an analysis of the functions. This is because the functions are most often clearly related to the purposes of an organization. This means that they are easier for the user to discuss. The existing recording system is likely to have had a determining effect on the users' perception of the entities. Care should be taken not to carry over into the new system limitations imposed by the old one.

A hierarchy chart is used to show the major functions and the way in which these consist of other simpler functions. This breakdown of the whole into its parts is known as top-down decomposition. (The term stepwise refinement is sometimes used with similar meaning.) This is an important technique in systems analysis and design.

Figure 4.1 shows the function chart for the Video Shop. Figure 4.2 is for the Air Freight company, and fig 4.3 for the Professional Institute.

Notice how in the Video Shop, there are three major functions —administration, dealing with suppliers, and dealing with customers. It is hardly surprising, in a shop, to find that the largest group of functions is the 'Customer Service'. However, it is not always true that the functions which occupy most time also occupy most space on the chart. Some simple functions may be repeated many times.

In the Video Shop there are actually several different functions associated with serving customers. The first sub-division is between Video Services and the rest (labelled non-video services). There are four functions associated with Video Services—returning, hiring and selling videos and registering new members of the video hiring club. Finally each of these is broken down into simple functions.

Readers familiar with structured programming will see the resemblance with the function chart. We have not created a program design yet. Many of the functions may never be computerized. Even where they are the program structures may have to reflect technicalities as well as the logical function structure.

4.3 ENTITY MODELS

Once agreement has been reached on the best function decomposition, then the next stage is to determine what information is associated with each one. This consists of modelling the entities, relating entities and functions, and finally charting entity life cycles.

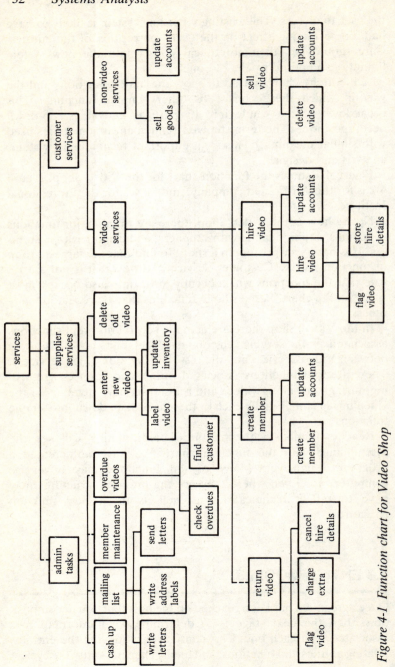

Figure 4-1 Function chart for Video Shop

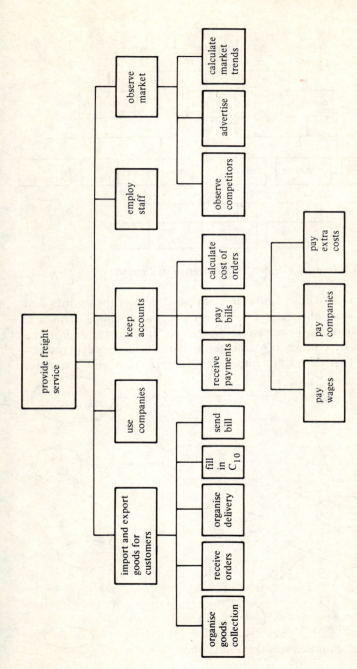

Figure 4-2 Air Freight: functional model

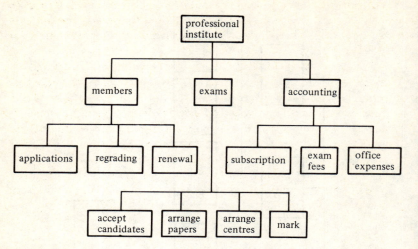

Figure 4-3 Function chart for Professional Institute

You can make an entity diagram which shows the entities and their relationships (e.g. fig. 4.4 for the Video Shop). This is a slightly theoretical exercise as there is room for judgement in the selection of entities (i.e. the things you want to keep records about). Two things are important: completeness and economy.

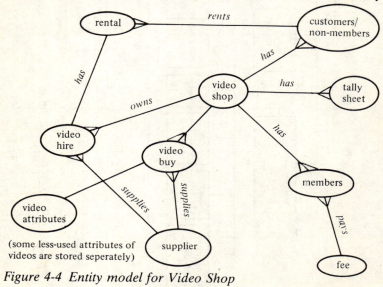

(some less-used attributes of videos are stored seperately)

Figure 4-4 Entity model for Video Shop

The best test of the proposed entity model is to check that each
record is used for at least one function and that each function uses
at least one record. Figure 4.5 gives the entity function matrix for
the Video Shop.

Finally, it is very important to know how long an entity is likely
to remain in the system because this will have considerable
repercussions when the physical database is set up. If a lot of the
entities stay in the system for a long time then more space is
needed, whereas if there are many short-lived entities then there
might be a backup problem. Entities can change into other
entities, e.g. for the Professional Institute the applicants could
become student members when they paid their fees. When they
passed their examinations they became members. They could even
become lapsed members if they failed to pay their fees.

4.4 ENTITY LIFE CYCLES

Some records will remain in the computer for a long time, others
will not. It is important to know what event will trigger the
creation of a new record of each type, the event that will change it

	video services	non-video services	sell goods	update accounts	return video	create member	hire video	sell video	enter new video				
rental					X		X						
video shop	X	X	X	X	X	X	X	X					
customer	X		X										
member	X		X		X	X	X	X					
video hire					X		X						
video buy	X							X					
supplier									X				
fee					X	X							

Figure 4-5 Entity function matrix for Video Shop

to another type, and what will finally cause it to be deleted. Figure 4.6 shows the complex life cycle of the records associated with someone applying for membership of a professional institute. An applicant can become a student member who may become a full member on passing the examinations and may even become a lapsed member on failing to renew subscription.

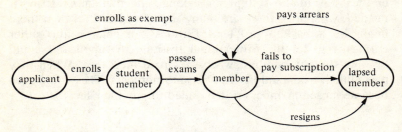

Figure 4-6 Entity life cycle for member of Professional Institute

Some entity types will be, by their very nature, short lived. Records of payments outstanding or goods on order should not need to be kept for more than a few weeks until they are settled. Some records will only last for a day, e.g. a bakery company may collect orders during the day and keep a record of goods to be despatched that night. You may even have transitory records, such as an enquiry in process which is deleted as soon as the answer is known.

The length of time that particular records might need to be kept in the computer vitally affects the storage space required to hold them. Some records, like tax returns or audits, will have to be kept by the organization for a long time. Often it will be enough to keep these in an archive somewhere (e.g. on microfilm, paper, or magnetic tape) where they are not immediately available for computer processing, but can be retrieved for people to look at. For many purposes it is not worth keeping detailed records even for reference, but you might want old records to be subsumed into general statistics. For example you may not need to know exactly who bought all the goods that you have sold, but you will want to know how many of different things have been sold, and possibly keep a profile of the types of customers who buy different sorts of things.

4.5 ARCHIVING OF INFORMATION

There are some types of information that have to be kept current for many years, most notably insurance policies which may need to be kept for the best part of a lifetime. This is quite exceptional. There is usually a point at which records cease to be logically required by the system, e.g. members that have left, examinations that have been taken. These can obviously be relegated to an archive of some sort. For large volumes this could well be microfilm. This can be generated straight from the computer without the need to print anything. The equipment is expensive but specialist firms will undertake it for you. You can archive data for a year or so on magnetic tape, but this is subject to deterioration eventually, and if you keep it for too long there is no guarantee that some new computer will still be able to read it. However, tape is very suitable for accumulating records for the year-end audit etc. For small companies archives of important material can be kept on paper.

Archiving is partly logical but it is also partly practical. There is obviously a limit on how much data can be stored in any one computer. If files are allowed to creep too close to the maximum capacity, there will be slowing down of operations as the operating system hunts for space to store new records. Moreover, there is always the possibility that data will be lost if the space suddenly runs out. This problem must be studied in detail after a computer system has been chosen, but general thought should be given to the importance of keeping different types of records within the system and deciding:

How long a record *has* to be kept current.
How long it would be desirable to keep it, and how much that would be worth.
What sort of statistics and historical archives need to be kept.

4.6 DATA FLOW DIAGRAMS

These relate functions and events to the flow of data through the system. Data flows from left to right across the diagram through the functions. The functions are shown in boxes. Events come in from the top. Confusion can arise with words like 'tally' (fig. 4.7) because it can refer to an entity (the tally sheet), a function (to tally up), and an event (taking tally).

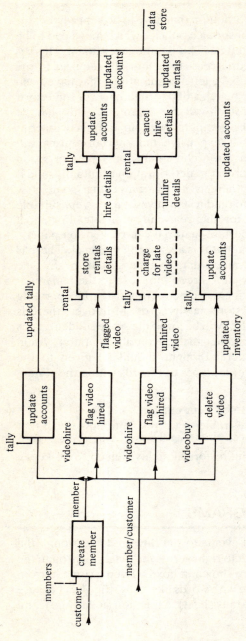

Figure 4-7 Video Services data flow diagram

4.7 THE BOUNDARIES OF THE SYSTEM

When embarking on a systems analysis there are two pitfalls to be avoided. The first is to concentrate too narrowly on the task in hand without considering how it relates to the other activities going on around it. The other is to be so carried away with the relationship between one activity and another that you finish up wanting to rearrange the whole organization.

The only way to avoid this problem is for the problem owner and problem solver to get together and discuss potential costs and benefits of considering a larger or smaller definition of the problem.

The most obvious reason for wanting to extend the scope of a system is where there is a lot of data shared between the part of the system under consideration and other areas of the organization. It should be clear from the entity model whether significant numbers of the entities are involved in other organizational activities.

Similarly one set of functions may appear to be closely related to another, or the same events may trigger actions both inside the proposed system and outside it.

The systems analyst can propose different alternatives for the extent of the system—perhaps a simple system to handle the kernel, the system as originally proposed, an extended system to integrate some closely related activities. The various costs and timescales for the three would have to be estimated. Then a commercial judgement must be made firstly as to whether the benefits are greater than the costs (in money, effort and disruption) and secondly as to which alternative represents the best investment.

Indeed there may be constraints on the extent of the system. There may be an absolute limit on time or money. There may be practical, personal or political reasons why the scope cannot be extended; you can't go computerizing someone else's records without their agreement!

Thus the information model is finalised by drawing a boundary around that part of it which is to be considered in the current study.

4.8 SUMMARY

Stage 2 of the systems analysis and design is the creation of a model of the information system. This is a logical model with no thought for HOW the system is to be implemented, only WHAT it is to be.

The three components of the model are functions, entities, and events:

1. Functions are analysed by top-down decomposition to produce a function hierarchy chart. In this the major functions are shown at the top of the chart. These are then broken down into their major components which are further broken down until single tasks are shown at the bottom.

2. Entities are things you keep records about. There is room for juggling between things that have their own records and those which appear as attributes on other records. You can chart the various entities and their relationship to one another.

3. Events occur within the system or outside it, and they trigger functions.

To ensure the completeness of the model matrices should be prepared relating entities to functions, entities to events, events to functions. Data flow diagrams relate events to the flow of information through functions. Entity life cycles show how long any particular entity is likely to remain in the system.

An information model is completed by drawing a boundary around what is to be taken as the information system for current consideration.

5
Socio-Technical Design

There is no point in creating an information system which does not fit into the running of the organization or which staff are unwilling or unable to use it effectively. Thus it is essential to consider the way in which people carry out their work and the way in which the new system can best be fitted into it.

Socio-technical design includes job design, specification of the way the system will be used, and decisions about staffing and training requirements. Most of these tasks are for the problem owner and the systems users. It also includes consideration of the hardware configuration and the way work is progressed through the computer, because this is the most critical factor in determining how the work is organized. What the analyst has to do is to lay out the various alternatives to permit sensible decisions to be taken.

The process starts by agreeing social objectives and technical objectives. 'Technical' is used here in the sense of what the machine has to do. Then there are various social alternatives describing different ways of organizing the work, i.e. you can assign different tasks to people, and group them in different ways. Similarly there are technical alternatives of how this may be achieved. The machine can do different parts of the work, or do it in different ways. All the combinations of social and technical alternatives can then be compared and the best selected.

If the system is a large one involving many people, then arrangements must be made to ensure that they have the chance to participate in the decisions which affect them. The people who are actually dealing with customers, processing sales, checking medical records, etc., are the ones best able to clarify the requirements for that part of the system with which they are involved. They can't tell you what machine you ought to buy or how to balance their requirements against those of other users. Different people will have different parts to play in the decision making process. Figure 5.1 shows the socio-technical design for the Video Shop.

5.1 SOCIAL OBJECTIVES

There is a variety of possible social objectives, some of which will be contradictory. When you are feeling bored, you will tend to ask for more job variety, but when you are confused by having to do many different things you will ask for life to be made simpler. No one will ever be perfectly happy with their job all the time. You just cannot achieve everything, and what you want to achieve will depend on the situation in which you find yourself. Various factors ought to be considered:

● Are the existing staff going to use the new system? For the Video Shop the Air/Freight company, and the Professional Institute, the system was to be used by existing staff and therefore had to be acceptable to them and to fit into their work. Are there any problems which need to be alleviated, e.g. boredom and high staff turnover from excessively repetitive work? In some large organizations the clerical work is as routine and repetitive as any production line. However most of the problems were incurred by the staff not able to get at information they needed, rather than any difficulties they had over working conditions.

● To what extent will the system affect the work of different users? In the Video Shop for example, it was necessary to assure the accountant that his work would be possible even though it might not be an ideal system for him.

● Are there specific objectives, such as improving the effectiveness of key personnel? In a merchant bank for example, all information processing activity is geared to making sure that the dealers get the best possible support because the whole operation is dependent on their dealing. The bank could go out of business if they lent money at the wrong rate of interest.

● What emphasis is to be given to general objectives such as 'improving job satisfaction' or 'being acceptable to all employees'? Where there are large numbers of staff doing routine clerical jobs which are about to be computerized, these factors could be critical to success. In smaller companies there is often more job variety

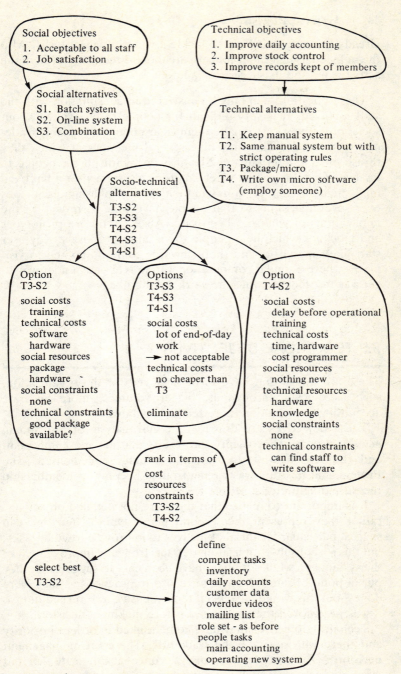

Figure 5-1 Socio-technical design for the Video Shop

already because staff have to tackle a wider range of tasks. Also the individuals in a small concern are more directly involved in the actual work and its success or failure.

These questions can only be answered in consultation with the staff involved. It is very difficult for people to be explicit about social objectives, or to make them clear enough, to enable trade-offs to be made. This may take several meetings at which they thrash out the alternatives. Also it is very difficult for people to imagine what it is going to feel like using a system after it has been implemented. Deliberations can be helped by phasing discussions over several sessions so that people have time to reflect on the implications of what has been said. Also helpful are prototypes and mockups which help people to visualize proposals. Key user staff like supervisors and shop stewards could be invited to visit places where a similar computer operation is already under way to get a better feel for what the work will be like.

5.2 TECHNICAL OBJECTIVES

These are things that have to be achieved for the organization, such as quicker throughput, bigger volumes, more accurate accounting, information more readily available.

These objectives are drawn from the analysis of the human activity system. They signify the results that are most important and the reason why the system is being implemented. The Professional Institute was having trouble keeping its membership records straight etc.

It is important to realize that these are not detailed objectives, rather they represent the 'ground plan'. Quite often one can suggest particular technical objectives to improve particular tasks, but this sort of sub-optimization detracts from any improvement of the system as a whole. It might have been easy to suggest that they put in point-of-sale terminals to record transactions as they were done—but this would have been too expensive as it turned out, so it was just as well that we waited before making a decision.

Technical objectives may have to be ranked in order of priority and there will inevitably be trade-offs. These are management decisions. The duty of the analyst, once again, is to lay out alternatives, and to spell out implications.

5.3 SOCIAL ALTERNATIVES

These are different ways of organizing the work. There are almost as many possibilities as there are offices, but here are some of the main alternatives:

Is the work all going to be collected into a daily batch and run in one go?

Are all the staff going to have terminals for data entry?

Are some/all of the tasks to be performed by specialized operators?

Are some/all of the tasks to be performed by untrained users (e.g. public use of a computerized library catalogue)?

The picture is actually going to be more complicated than this because different parts of the work can be done in different ways. For example video hirings could be recorded as they happen, but all new stock entries could be done as a batch.

5.4 TECHNICAL ALTERNATIVES

These are different ways of organizing work on a computer and of splitting work between computer and manual operations. Different alternatives might seem appropriate in different situations, and in any particular situation some of the possible alternatives may seem technically easier to implement than others.

One alternative that has always to be borne in mind is to keep things exactly as they are, or to insist that they be run properly, in the way everyone agreed they were supposed to be run. The analysis of a perceived problem may lead to the discovery that the problem actually lay elsewhere, particularly in careless operation, or else that the problem was inherent in the situation and that the effort and expense of implementing a new system would not be justified by the benefits. This alternative was given serious consideration in the Video Shop because the paper system ought to have given adequate information. After observation it became clear that apparently careless documentation was caused by sales assistants putting it aside to perform their first duty, which was to serve the customers.

Other alternatives will depend partly on the size and nature of the problem. For a large system the alternatives might be:

a large central computer,
a distributed system, i.e. several computers each handling part of the work but interconnected in some way,
either of those with intelligent terminals to permit on-the-desk processing as well as having access to shared files,
batch processing, real-time operation, or a mixture.

Note that the decision of what sort of system is to be installed must be made by management. The analyst must give suitable advice on the likely cost, implementation time, operating costs and considerations, etc. However, the technical factors are not the only ones influencing the decision, which is why it is so important to go on to the next stage.

5.5 SOCIO-TECHNICAL ALTERNATIVES

Here you simply take all the social alternatives and all the technical alternatives and try out all the combinations. Some of the combinations will be wildly impractical for technical, financial, or social reasons, and can therefore be ruled out. This should leave a few realistic alternatives. These can be studied in more detail to yield costings, approximate lead times, likely social effects in restructuring people's jobs, the extent to which they might improve the company's operations, etc.

The elucidation of the alternatives will be a more or less detailed exercise depending on the size, novelty, and complexity of the proposed systems. For a microcomputer system, the salesman could probably lay out the alternatives for a straightforward system with the aid of the brochures on his counter. For a large system such as for a government department, it would take a team of experts months to lay out a rough plan for even one of the alternatives.

A cost-benefit analysis of the alternatives should be carried out to include not just money costs but also social and technical costs and benefits. Social costs might be the necessity for redundancies, the creation of jobs which are more rigid or boring, the disruption in setting up a new system, the amount of retraining required. Thus a decision can finally be taken to select an alternative for further work which is expected to give the best technical and social return for the money and disruption. This is a commitment to a certain type of system (e.g. a large central computer with many terminals). It is not a selection of a particular make of computer. A whole lot more detail has to be worked out before you are even in a position to invite the suppliers to propose the equipment you might want to look at.

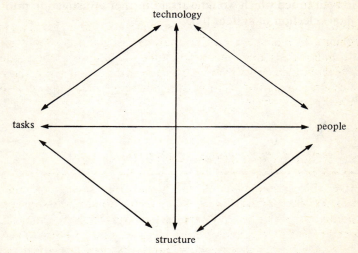

Figure 5-2 All the inter-relationships in the socio-technical system

5.6 SUMMARY

A computer system must be integrated into the working lives of the people using it. The organization must spell out its social objectives, such as:

reduce staffing, or else maintain present staff,
reduce boredom, provide job variety,
be acceptable to existing staff,

and so on.

There will be various alternative methods of achieving these by organizing the work in different ways. For example:

collecting work into batches to be run by specialist operatives,
all staff having their own terminals and doing their own processing.

Technical objectives for the system are the functions it must perform. There will be technical alternatives, i.e. different types of computer and/or manual systems that could achieve these objectives.

Various combinations of social and technical alternatives can be tried out to see which would warrant further investigation prior to a final selection of system type.

6
Man-Machine Interface

Man-Machine Interface refers to the way in which an individual relates to the computer. The 'interface' may involve putting data in, having queries answered, creating or reading reports, documents etc., initiating actions, performing security and backup functions, and so on. In fact anything where information or instructions pass between a human being and a computer.

The term 'man-machine interface' has been criticized on several counts. Firstly it is more often *women* who actually use the computer, particularly for data entry, and this has offended those who are already concerned that women are under-rated in computing. Of course, 'man' is intended in the generic sense. A second complaint is that it seems to put the machine on the same footing as the person as if they were equals having a conversation. The term 'human factors' has been suggested as an alternative and is used by some authors.

There are two strands in MMI research. One is concerned with the technicalities of speech recognition, voice synthesis, pattern recognition, etc. These are technologies which will enable us to communicate with the computer other than through the keyboard. This is the province of the computer scientists and engineers. The systems analyst is concerned with the other aspect of MMI and this is the logic of the interaction, i.e. *what* information needs to be exchanged and *what* conversation needs to take place in order to effect that exchange. The aim is to make the conversation as clear and precise as the computer needs it to be, and yet as natural as possible for the benefit of the humans.

The computer will never be able to obey commands which are incomplete or ambiguous, any more than a human subordinate could do, at least without making assumptions about what the command should have been.

These assumptions can be programmed into the computer if you want them to be. For example 'Recall it' would be taken to mean

'Recall the document I was last working on'. 'Put it on the desk' is an ambiguous command because it doesn't say what is to be put or which desk it is to be put on. However if someone comes up to your desk while you are speaking on the telephone and offers to show you a report, then the context makes it quite clear that what you meant was 'Put the report that you are holding on my desk'. Because the situation is a familiar one we all fill in the missing information without even noticing.

When you are getting information from people, you start off by treating them as functionaries, i.e. you have a stereotype of how to talk to a librarian or railway bookings clerk. If you deal repeatedly with the same people you start treating them as individuals. Unless the two of you really imitate each other, this individual mode of conversation is likely to be more productive in terms of the amount of information conveyed. This is summed up by saying that you 'understand each other'.

Computers still cannot work in fluent English (or Japanese etc.) although considerable advances in our own understanding of the logical rules underlying language are facilitating the use of simple English commands. Voice input is still limited—usually to single command words uttered by a user whose speech has already been recorded for comparison. Voice output is readily available, although it is primitive and without expression.

Much of the work being done on MMI has been psychological study of how the user understands what the computer is doing. This has shown up the user's need for a 'mental map' of what is going on inside the computer. This map enables the user to understand the available commands rather than learning them by rote. It does not matter so much whether that map corresponds to what the programmer thinks is happening inside the computer (he has his own mental map), as whether the map is logically consistent and coherent with the user's view of the real world functions being performed. Staff training on the computer should always take into account the user's need to get a feel for what is going on 'inside the system'.

6.1 WHICH STAFF WILL USE WHICH PROGRAM INTERFACES?

This question is central to the design of man-machine interfaces.

It is not practicable to tailor the computer system to individuals, so the next best thing is to pick out those categories of staff who will be using each interface and whose preferred modes of working are likely to be similar. For example accounts clerks will have mental maps of the company's operations which feature invoices, debits and credits, payments, receipts etc. They will want to enter individual items and check details carefully. Managers on the other hand are likely to want to talk about targets, aggregates, trends, etc. They will want to be able to manipulate large volumes of data quickly and may well be uncomfortable using a keyboard. Obviously the sorts of interfaces you build for these different categories of staff ought to reflect those differences.

6.2 DIALOGUE DESIGN

Perhaps the simplest way to analyse the interface requirements is to think about the dialogue that has to take place between user and computer. We need not yet worry about the form of the dialogue, only its content.

A dialogue may be very simple, such as:

'Has member 4371 paid his subscription?'
'Yes'.

When you look at this, though, there is actually more to it, for instance how did the user know the membership number? This may actually call for a preliminary dialogue, at least for those members who don't quote the membership number on their correspondence. There is also the question of what ought to happen if the computer can't find a membership record for number 4371. This may require some subsequent dialogue to find the right record. There are other possibilities to be considered if the dialogue is to be completely clear, such as what happens if the member has paid some money, but not the right amount, etc.

To provide a customer with a video involves several different dialogue exchanges:

a browse through what is available to make a selection,
determining whether a required video is actually in stock,
possibly the acknowledgement of a purchase,
possibly the confirmation (or creation) of membership, followed by the acknowledgement of a hiring.

Closer analysis of the details of each transaction might disclose that it is a complex of several individual transactions. This is especially true where errors have to be dealt with. What is surprising is how many places an error or exception can occur when trying to do something on a computer. For example if someone tried to order a video that did not appear to be on file, you must know what to do about it. You might want to try another number, create a new record, find out what went wrong, or whatever. The programmer should not be left to his own imagination in dealing with these errors and exceptions. It is often said that 80 per cent of the instructions in a computer program are to deal with things that might go wrong. The instructions for doing the straightforward task account for the other 20 per cent.

Sometimes it is quite difficult for users to work out in advance all the things that might go wrong during a dialogue and then what to do about it in each case. This is because these errors crop up so rarely that people don't often think about how to guard against them. Also, many errors can be spotted by 'common sense'. What this means is that people have learned by experience what should be done in certain circumstances, and how to recognize problems. The analyst has to help them spell out in the dialogue the results of that experience. Scientists are working on the theory of computers that can learn for themselves, but for now we have to make all the instructions absolutely explicit.

Very often the simplest way to deal with unforseen errors will be to abort that particular transaction and try again from the beginning. It is a well-known fact that we make many times more errors trying to correct mistakes than we would do normally, so it is often quicker to begin again. However, if you have put a lot of data into the computer, it should ask you whether you want to change part of it rather than make you scrap the whole thing.

6.3 MATCHING THE DIALOGUE TO THE USER

We have already mentioned the fact that different sorts of users will want to talk about different sorts of things. For example accounts clerks might want to talk about debits and credits while managers might want to talk about sales trends. In addition to this users will have a different way of constructing the dialogue according to such things as their familiarity with the system. This doesn't just apply to computer systems. A reluctant do-it-yourself plumber who needs to mend a sink might well have a rambling dialogue with the builder's merchant while he explains his predicament and tries to describe the part that he needs. He may resort to drawing a diagram or pointing to similar items on the shelves. He will probably be sent back to measure the exact diameter of the pipe etc. before he is actually able to complete his transaction. He may even have had to try several shops before finding one that kept the right sort of stocks. The professional plumber will go straight to the shop and give a precise specification of what he wanted, which the assistant will be able to recognize immediately. The same part in a well-organized warehouse could be ordered simply by quoting a part number.

This represents three different levels of dialogue:

● The complete beginner who needs to be prompted for the necessary information and whose instructions must be carefully checked to see if they make sense.

● The regular user who knows how to get things done and who wants to do them quickly and efficiently.

● The expert who knows how the inside of the system works as well as knowing how to operate it.

Each of these wants to conduct the dialogue in their own way. If a dialogue is designed for complete beginners then it will include lots of explanations which would slow down the regular user. This would reduce the number of operations that the regular user could perform, and would probably be very irritating. It would be rather like having to say 'Hello, how are you? Isn't it a nice day. I hope your cold is better' every time you spoke to the person at the next desk. The 'expert' level of dialogue should be reserved for the people who have responsibility for the system and who can be trusted to use these powerful commands properly, e.g. for repairing errors in the database.

6.4 METHODS OF IMPLEMENTING A DIALOGUE

The earliest form of dialogue was the 'command', where you simply typed in a command and the computer was expected to obey it, e.g. on the BBC microcomputer CHAIN "" means load in the next program on the tape and then start running it. To make sure that even the most unskilled users know how to make the program work the loading instructions are often printed on the tape label. That is fine if you kow what the command was and you were sure that you had typed it correctly. Even skilled operators have problems with this mode of operation now because there are so many commands available on today's computers that you can easily forget exactly which one you wanted to use.

It is also important that you know the 'command structure', that is at which points you can use which commands. 'Right Turn' commands can only be issued to soldiers when they are standing to attention. If you want them to turn right when they are marching you have to say 'Right Wheel'. Computer commands are similar, e.g. the CHAIN command can only work if your computer is already operating the BASIC language. If it had been operating the word processor you would have had to change to BASIC first. On the best systems you can often type one command string with the effect of

```
*BASIC
CHAIN "PROG"
```

to invoke both operations.

The use of command strings is obviously unsuitable for the user, although it can be ameliorated by the use of a 'help' command which can explain the other commands to you. The next stage is to display the available commands in the form of a menu and invite the user to select the one required. This avoids the necessity for remembering the actual commands, but you still have to know the command structure. This is because you still have to know how to get from the menu that is on the screen at the time to the one which offers you the procedure that you actually want to use. These menus are often arranged in a hierarchy which resembles the function hierarchy of the information model. This is a useful method for situations which are inherently hierarchical in structure so that you can get to the right set of commands and run through a whole batch of similar work.

It is quite possible for the computer to take the initiative in a dialogue, especially when collecting data. The computer can prompt you by displaying questions on the screen which you have to answer. This is helpful for the novice level of dialogue because it can provide an explanation of what information is required and lead the user gently in the right direction. Some doctors are using computer-based dialogues like this to gather some of the information they need for their medical histories. These dialogues can be structured so that different questions are asked depending on the answers that have already been given.

'Do you have pain?'
'Yes'.
'Is the pain in your
head,
chest,
back,
stomach,
etc?'

It can also prompt you by displaying a form on the screen for you to fill in. This is commonly used for data input where the original data was already formalized. At its most fool-proof this can consist of operators filling in the blanks on the screen with data from the same boxes on the paper form in front of them. This also makes for a very boring job. This is a style of dialogue for the regular user.

Much has been made of the possibility of talking to a computer in your own language. This is really just another way of issuing a command, except that you go through an initial dialogue where the computer looks through a dictionary and a simple grammar to see whether it can interpret your English into formalized English phrases which corresponds to the commands it already understood. It displays back to you what it thinks you have asked for, such as:

'I want a printout of which salesmen earned the biggest bonuses this year'

'PRINT SALESMAN-NAME WHERE BONUS???'

'I DO NOT UNDERSTAND BIGGEST BONUSES'

The computer has recognized the words 'print' and 'salesman' and 'bonus' from its vocabulary and arranged them into its own query language. However it is not quite sure what to do about the bonus. The manager needs to have another go.

'Print salesman-name where bonus greater than £5,000'.

'PRINT SALESMAN-NAME WHERE BONUS GREATER THAN £5,000'.

The computer now understands the command and can execute it. What is most interesting to observe is that the naïve users trying to use a natural language interface all learn the computer's own version of the query language much more quickly than the computer seems able to learn English. Thus the beginner soon becomes able to operate at the regular-user level of dialogue.

6.5 COMMUNICATING DEVICES

We have not yet said what equipment is to be used to enter the commands or the data into the computer. The assumption has rather been that it is a keyboard, just because this has always been the convention. Voice input is only available for the simplest commands. If you think about it voice input is probably not a very good way of inputting solid amounts of data. This is because a room full of operators would all be talking into their machines at the same time. The machines would get noise from each other, and the office would be a babble. The computer would probably also be trying to decipher commands like 'Hello Mary, did you have a nice weekend'. However, it obviously is a very good idea for situations where the operator cannot use his hands. Pilots can have voice input devices in their cockpits to recognize specific commands. Some cars have them. For the handicapped they can be a lifeline.

Where there are very large amounts of very formal data, this can often be input better through a device like an optical character reader (which can recognize typescript and careful print) or an optical mark reader which can recognize marks on a form. The latter is used notably for meter reading and checking pools coupons. Where a form has to be sent out and then returned it is possible to use magnetic ink character recognition (like the letters at the bottom of a bank cheque).

The keyboard is still a very efficient device for data entry where regular operators are going to work because high speeds and good accuracy can be achieved with practice. There is also a high level of experience among office personnel in the use of keyboards so staff know how to use them and are confident dealing with them. New designs of keyboard are being brought out which can increase throughput by as much as 30 per cent in even the most skilled qwerty operator. However, there is obviously an upper limit.

Where non-skilled staff or clients are going to be operating the computer, then the amount of keyboarding can be reduced by judicious use of menus, so that most of the commands can be issued with only a single key stroke.

A device which has caught the popular imagination recently is the 'mouse'. This is a small box which can be held in the hand and rolled around on a table top. It is connected to the computer and operates a pointer on the computer screen. It is used with a very high resolution screen which appears to have documents sitting on it. It also has symbols like filing cabinets and waste paper baskets. By pressing a button on the mouse when pointing at a document and then moving to the wastepaper basket and pressing the button again, you indicate that you wish to throw away the document. This has two advantages over more conventional ways of giving the instructions—you do not need to know how to type and you do not have to understand any of the languages of the computer. Perhaps this is not entirely true because a visual language is a language of a sorts. Symbols work very well for simple things but how well they can convey complex meanings precisely enough for the computer has yet to be shown. One practical problem with the mouse is that you also have to have a clear space on your desk for the mouse to move on!

Computer output is also part of the dialogue. So far the only responses we have mentioned are words on the screen. We could also have words or figures on paper. Modern VDU screens give the possibility of producing graphical output—converting figures into graphs is often a better way of describing them. Using dot-matrix or ink jet printing you can get printed versions of any graphics from the screen. Using a graph plotter or a high resolution screen the output could be a complex diagram such as a blueprint. Voice output is easier to achieve than voice input and all sorts of other sounds are possible. Sounds obviously have their uses but few people would want to encourage any additional noise in the office, and in some environments e.g. hotels, libraries and hospitals, noise would be shunned.

6.6 SUMMARY

People get the most information out of a dialogue when they know who they are talking to. The nearest one can get to this on a computer is for the dialogues necesssary to perform a particular transaction to be carefully suited to the natural style of the person who will be using it. It is not practicable to design dialogues for each individual user but it should be done for each group of users who are likely to want to do the same thing in the same sort of way.

The first stage is therefore to identify these groups of users and analyse the transactions they will be performing in terms of the exchanges that this will involve between them and the computer. Particular thought should be given to what should happen if the dialogue does not go according to plan because of some error or exceptional situation.

Different levels of user can be distinguished according to their level of familiarity with the computer and the information contained in it:

beginners who need to be guided through the dialogue,
regular users who should be allowed to proceed quickly straight through it,
experts who are going to manipulate the computer system itself (often these will be inside the DP department).

Keyboard and screen are the most common devices for running these dialogues. Beginners can be helped by reducing the keyboarding to a minimum such as selecting items from a menu. The keyboard can be supplemented with a mouse. Greater throughput for high volumes of routine work can be achieved by the use of optical character or mark readers or even magnetic ink readers. Other input/output devices can be incorporated for graphics, print or possibly voice.

7
Database Design

The word 'database' is used in two senses. Depending on the context it can refer to the stored information or to the 'logical map' used to organize the way different records are stored and related to each other. The analyst is concerned with creating that 'logical map'. The user who says 'the figures are in the database' is referring primarily to the stored information itself.

In order to manage a database one would normally purchase a DataBase Management System (DBMS). Each DBMS works on a particular logical data structure, i.e. it organizes its records in a certain way. It provides all necessary facilities for specifying the content of records and the way in which different types of records are inter-related. It also provides all the technical facilities for ensuring data integrity etc.

The human brain is amazingly flexible in the way that it cross-references information. We do not normally have to think about how we jump from a face to the person's name or all the other bits of information we recall about him. We have to be much more careful about information which is stored on paper, although we can often get away with writing things 'on the back of an envelope'. Even when we use preprinted forms for structuring the information, we can still use some flexibility in writing notes across the top of the page or whatever. Such papers tend to be filed rigidly (for example in alphabetical order) or else to be in a totally unstructured heap on your desk. You may be lucky and know exactly where in the file to find the thing you want, or you may resign yourself to looking at every piece of paper in turn until you find what you're looking for. There is also the hope that the people who wrote the information will remember how to retrieve it, e.g. 'Could you let me have the file on that fellow who was here yesterday making such a terrible fuss'.

Computerized information systems are much more reliable than the human brain, or the notes on the backs of envelopes. They are much more versatile than paper ones in the way information can be retrieved or processed. However there is an absolute limit to what it can do. You can only get out what you put in. That goes not only for data items but also for relationships between them, such as in a hospital database you might have records of doctors and patients but unless you indicated the relationship 'Dr. A treated Mr B' you would never be able to find out who treated whom.

This chapter shows some of the ways in which data records can be organized within a computer. The most versatile ones are usually also the most difficult (and expensive) to implement. Different methods of organization may be better for particular applications because they are better representations of the actual relationships in the data to be shared. The actual relationships were drawn out in the entity diagram in Chapter 4.

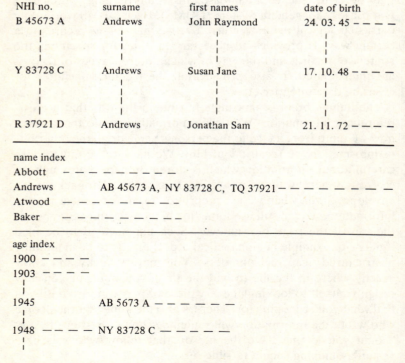

Figure 7-1 A multi-index file of doctors' records

7.1 INDEXED FILES

Where a data file is kept it is often necessary to access records according to different criteria such as a doctor keeping a record of patients. Each record can be uniquely identified by the NHS number, and this is known as the *primary key*. Normally the doctor would want to look up records by the patient's name, even though there is the possibility of two patients with the same name. 'Name' is thus a *secondary key*. Figure 7.1 gives an example of a multi-index file.

An index is kept for the primary key showing the exact location of the record. For each of the secondary keys an index is also kept. This shows the locations of all the records with each value. Thus there may be 'Smith' at records, 47, 603, 949 . . .

For a file which is not too large (tens of thousands of records) and which is relatively static, a multi-index database of this sort can be remarkably effective. It does rely on the fact that all the records have the same structure and that you know what the secondary keys are going to be. A good use would be in a library catalogue. The primary key would be the ISBN (International Standard Book Number). Secondary keys would be Author, Title, Dewey Number. For each book record there would be the same set of attributes (Title, Author, Publisher, date of acquisition etc.).

This style of DBMS would be appropriate for a situation where the entity diagram contained only one major entity that was to be computerized.

7.2 HIERARCHICAL DATA STRUCTURES AND NETWORKS

Many well-regulated office filing systems are based on a hierarchical structure, e.g. the Sales Records for a national firm might be organized as follows: (fig. 7.2).

The company has many sales offices.
Each sales office has many filing cabinets (one for each salesman).
Within each salesman's filing cabinet are many folders (one for each customer).
Within each customer folder are many order forms.

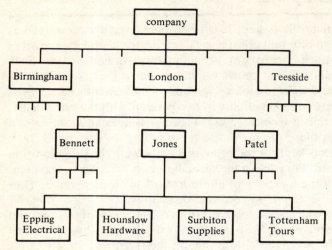

Figure 7-2 A hierarchical structure of sales records

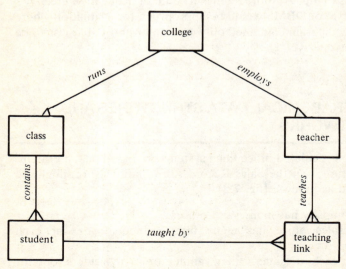

Figure 7-3 Student database showing the teaching link between student and teacher

This arrangement can readily be duplicated in the computer. Most hierarchical DBMSs offer an additional facility which cannot be represented on a paper system. This is the link between two parts of the hierarchical tree, e.g. fig. 7.3 shows a student database where teacher records are linked to those of the students they teach. Many different relationships could be represented between the same records. For example you can relate teachers to courses in their own departments, teachers to courses in other departments (and hence also to students in other departments). There can also be more than one link between teacher and student, such as teachers being responsible for the personal welfare of different students, as well as teaching some of them.

The relationships within sales offices are all 'one-to-many', with one sales office having many salesmen. (Many can be any number from zero upwards). The relationship between them is known as a *set*. The 'one' is referred to as the *owner* record and the 'many' as the *member*. The 'many' are shown by the crowsfeet on the entity diagram.

In a *hierarchical* DBMS the rule is that one owner can have many members. The relationship between students and teachers is 'many-to-many', thus each teacher will teach many different students and each student will be taught by many different teachers. Figure 7.2 shows a link record because it was actually implemented using a hierarchical DBMS—one student has many links, one teacher has many links, any particular link has two owners. One is the 'teaches' set and one is the 'taught by' set.

In a *network* the many-to-many relationship between teachers and students can be represented without having to define link records.

7.3 RELATIONAL DBMS

There are a great many limitations on what it is possible to do with a hierarchical or network database. You can only get information out within the structure. For example it is very easy to list customers by salesmen by regions, but there is no direct way to see who is buying what unless we extract all the order information and then sort it out in a separate program.

If you are collecting data which is going to be used for decision making in a changing world then you can be sure that you will be asking questions that the database wasn't designed to answer. This may be because the required relations were never built-in in the case of a hierarchical network. In a multi-index DBMS the problem is either that you cannot relate between enough different types of record, or that you did not have enough secondary keys.

The logical way out of this is to build into the database all the required data records and all possible relations. Computer technology has at last advanced to a point where it is possible to do this. It is called a *Relational* DBMS. Such systems may be more expensive, run on bigger computers, or run more slowly than the technically simpler type of DBMS. However this may be a price worth paying for the flexibility that they provide.

Implementing a relational DBMS requires a rigorous analysis of the data and the logical relationships between data items. Figure 7.4 shows part of a relationship database.

7.4 DISTRIBUTED DATABASES

An organization's data may be kept in several different places, even if these are only different offices in the same building. It may be sensible to set up several small local databases rather than one large centralized one. Reasons for doing this include:

> where most data is used locally with only occasional access to other databases,
> where the different databases have different types of records and different processes associated with them.

Distributed databases are easier to implement in one sense, in that each one is smaller, and therefore less complex. However, they do present problems such as:

> How do you protect data against unauthorized access by people at other sites using the telephone lines?
> There are technical problems associated with protecting one computer against the malfunctioning of another computer with which it shares data.

The only way to decide whether to use a distributed or centralized data base where there is more than one site, is to cost out the two alternatives and estimate the amount of traffic going between them. You will then be comparing the costs of equipment,

software, staffing, and overheads for one machine or several, and
the relative costs of telephone lines for all the different distances
involved between the individual user terminals and the computers,
and also between the computers of a distributed system. (Leased
telephone lines are charged by the kilometre.) You will also have
to compare the service that the two different configurations offer
the user.

LECTURER

LECTURER NO.	LECTURER NAME
00110	L. P. Antill
00143	W. A. Baker
00159	T. J. Hacker

STUDENT

STUDENT NO.	STUDENT NAME	COHORT	STUDENT ADDRESS
0017	Harsha Patel	BSc1	....
07371	John Smith	BSc2	....

LECTURER CLASS

LECTURER NO.	CLASS NO.
00110	Mon 930/463
00110	Mon 130/307
00143	Tue 930/466

CLASS

CLASS NO.	COHORT	SUBJECT
Mon 930/463	BSc1	COBOL
Mon 130/307	BSc2	Systems Analysis

Figure 7-4 Part of a relational DBMS

7.5 CHOOSING A DBMS

The analyst must obviously take into account the logical structure of the data to be stored, and also the type of questions that are going to be asked of that data. The suitable data structures have been discussed under the individual DBMS. However, there are other factors to take into account.

The most obvious of these factors is availability—Is there a DBMS of the required logical type available on a suitable machine? Whatever the advertising says, almost all microcomputer DBMSs are of the multi-index type.

The 'performance' of the DBMS with your volumes of data must be adequate for the numbers of transactions to be processed. The more secondary keys or relationships built into a database, the longer it takes to update a single item. This means that where a database is very 'dynamic', i.e. lots of insertions, amendments, and deletions you may have to choose between a faster machine or reducing the flexibility of response by leaving out some aspects of the original entity model.

An organization's database is one of its most valuable assets, so it is important that the DBMS chosen should have suitable built-in procedures for ensuring the security and integrity of the data. Backup copies of records should be made regularly in case the originals are destroyed or corrupted. There should also be methods of repairing the database where errors have crept in. These procedures may consist of 'winding back' the database to the point before an erroneous entry was made, scrapping the mistake and then re-running all the valid transactions since then. This requires that the system should keep a 'log' of all the entries so that they do not have to be retyped. After the database has been wound back the entries can be read from the log and the database brought up to date before users are allowed to type in any more entries.

Access to the database, or at least sensitive parts of it, should be limited to those personnel entitled to that information. This may be achieved by passwords or by limiting certain transactions to specific terminals in secure locations. Different degrees of sophistication can be provided by privacy systems. You may have to 'lock' a complete file, or some of the records, or certain types of records, or just particular fields within records. You may need to have different passwords for different areas of the database or for different types of transaction. There is often a different password for being allowed to read information, being allowed to modify it, and being allowed to delete particular records completely.

When you come to the details of database design you cannot divorce the logical considerations from practical ones such as the number of transactions to be processed, the arrangements for repairing data areas, and the question of who is to be allowed access to what data. You have also to take into account additional facilities offered by the DBMS for information retrieval without the necessity for coding programs.

7.6 QUERY LANGUAGES AND REPORT WRITERS

If your event/entity matrix shows a lot of occasions when you may need to look up all sorts of information stored on the database then you will not want to rely on the writing of special programs for each of these accesses. This is particularly true of the decision support system where it is often difficult to predict what questions would be needed in the future.

Many database management systems have associated with them a 'query language' which permits the user to frame questions in a straightforward way using commands which are easily learned or by selecting items from a menu on the screen. Some of the more sophisticated databases even offer 'natural language interface'. This accepts commands typed in English, looks up the words in its dictionary and applies simple grammatical rules to translate these into query language commands. These are displayed for the user to OK before being acted on in case there were any misunderstandings.

Many of the latest DBMSs permit the user to insert new records and update existing ones without having to get programs written. This is particularly true of micro and mini databases. Thus a user can design and operate a simple database management system without recourse to programmers. However, he is then responsible for all aspects of security and integrity. Many micro users have lost their databases because they underestimated the need for backup copies etc.

7.7 THE DATABASE ADMINISTRATOR

While we are on the subject of database design, it is worth pausing to consider the particular role that has emerged in large companies to deal with the database. This is the database administrator whose job it is to oversee the translation of the analyst's logical design into a physical reality on the computer and to ensure the security and integrity of that data in operation.

This job is largely one of co-ordination between various teams of programmers and between all the users to ensure that the data is used correctly, that errors and mishaps are corrected and that the database is secure against malicious or fraudulent access. Some large organizations have a person or department to do this job. For small systems the role would be undertaken within some broader responsibilities. What is clear is that, with increasing volumes of important company data in the computer, the safe-keeping of that data is a major responsibility.

7.8 SUMMARY

One of the major design tasks for an information system is the design of the database itself. In many ways this is a very technical task, but the analyst is responsible for the logical design of the system, leaving programmers and others to achieve the physical implementation of the design.

There are different types of DBMS—multi-index, hierarchical, network, and relational. These are quoted in approximate order of costs although there are obviously other factors affecting prices. The structure of the DBMS should be capable of supporting the entity model discussed in Chapter 4. The total number of records, the number of different record types and relationships between those different records types, and the volatility of the database (i.e. how frequently records are inserted and deleted) will all affect the choice of a particular DBMS.

Databases may be centralized into one computer or distributed between several linked computers. They may be supported by additional facilities such as query languages to enable ready retrieval of data. They should also have mechanisms for ensuring the privacy, security, and integrity of the data.

8
Software Specification and Selection

Chapters 3 to 6 discussed the ways in which the analyst can help the user to specify what information processing is required. This was in terms of the data to be kept, the functions to be performed on that data, and the various tasks and dialogues to be performed by the user.

Chapter 7 showed how the logical data structure could be implemented in a DBMS. The next stage is to specify requirements for software and to decide whether these requirements can be met by existing packages and which will require specially written software. If software is to be written then this will have to be specified in much more detail which is discussed in Chapter 10.

8.1 DRAWING THE SYSTEM BOUNDARY

It is important to remember that just because things *could* be computerized, doesn't mean they *should*. Tasks should be computerized only when the computer can do them more effectively. Thus the analyst must work out, in consultation with the users and programmers, which of the functions specified in the information model to computerize.

These should first of all be split into three groups:

(1)   those which must be computerized,
(2)   those which could usefully be computerized if the software was available,
(3)   those which are logically or technically too difficult to computerize, (often these will be functions requiring

human judgement and where you don't know enough to predict the questions in advance).

Just where you decide to draw the boundary will vary as different alternatives are considered; for example one alternative might be an existing software package which happens to implement some of the functions and not others. For the Air Freight company one of the main systems boundary questions revolved around the tariff data used for calculating import duties. There is a great deal of this data which is not very well structured, and gets updated irregularly. It consists of a chapter number, a title or description, and then sub-descriptions at various levels. At the lowest level each sub-description has associated with it the data used in the calculations for duty. It was decided that the advantages of having the computer search through the 500 pages of tariff data were so great that it was worth trying to computerize it. Later in the chapter we will consider ways of evaluating these alternatives.

8.2 STATEMENT OF LOGICAL REQUIREMENTS

This takes the proposed system boundary and collects together all the entities, functions, and events included within the boundary. This is the statement of *what* the system must do. The socio-technical design tells us *how* the work must be done, such as interactively, in batches, or whatever. From the man-machine interface design we have a specification of the various dialogues that have to take place during the performance of various functions. We also know what type of database we require (hierarchical, relational etc.). Along with all these requirements, we also know something about the organization in which it is to be used. The type of system that would be suitable in a small family business might not be appropriate in a professional partnership even if it is doing the same job.

8.3 TURNING REQUIREMENTS INTO PROGRAM SPECIFICATIONS

A program is an instruction to the computer to perform functions on data. Thus the first stage of specifying a program is selecting out the part of the function chart that is to be implemented on that particular program. Thus the analyst must look at the functions to see which ones need to be performed together. In addition to the functions that have to be performed in order to conduct the user's business, there will be functions imposed by the fact that the system has been computerized (i.e. there will certainly be functions like: 'Take back-up copy of day's transactions'). These will be discussed further in Chapter 9.

The way in which the work is to be organized in a socio-technical sense will also generate functions. For example there was some doubt in the Air Freight system whether the operator would want to print out Customs forms one at a time as the data was entered, or wait until a batch had been entered and then print them all out together. The latter is most efficient as there is a delay every time you stop to print. However, if somebody is waiting for a form you need to print it immediately. This created an additional function: 'Choose batch or single print'. It also meant that there was more than one print function ('Print a form' and 'Print a batch') in different places on the function chart, only one of which would be called up for a particular batch.

There are different techniques for specifying software and there are different levels of detail used in different companies. One set of techniques has been produced by the National Computer Centre. Printer and screen layout charts can be used to specify the exact format of displays and reports. There will be charts to show the exact structure of databases and files. These will show the attribute fields for each record and whether they are alphabetic, numeric, alphanumeric, or logical field (i.e. whether the answer is true or false, such as whether membership subscription has been paid). For each field you must specify the number of characters it might contain, whether it will always have the same number of characters or whether it could vary. For a variable field, like a postal address, you need to indicate the delineation character that will be used to mark the end of it. Numeric fields need to have the maximum number of figures before and after the decimal point and also the number itself may be negative. A good programming system will also require 'range checks' to fields to make sure that they come within some permitted range of values. For example examination marks may be no greater than 100%, the date should never have more than 31 for the days, and 12 for the month.

One of the nicest ways of specifying the logic of a program is to use 'structured English'. This is a way of describing in words what

---

the program is to do without making any assumptions about the coding that will be used. Phrases are indented to show that they are part of a preceding phrase. This shows how the structure of the program is built up of major components which are themselves built up of minor components.

Example from the Video Shop
   receive customer request
     if request is for video hire
       enter video required
       check availability
       check membership
         if not member
           create member
       take cash
     else if request for video sale ...... etc.

This shows clearly which things you have to do if you are processing a video hiring and which you are to do otherwise. For a full program specification these would have to be broken down into further detail by additional phrases of 'structured English' at further levels of indentation.

Readers familiar with structured programming will see the similarity. This program design may be undertaken by a programmer working from the function chart.

8.4 RELATIONSHIPS BETWEEN PROGRAMS AND DATA ITEMS

A particular program may create data records, or certain types may alter some or all of the fields in a second type, or may delete certain types of record. Some programs will be allowed to read some or all of the data of a particular type. Certain programs will be forbidden any sort of access to certain types of record.

Some of these access conditions are logically necessary, such as the program to register a new member of the Professional Institute must be able to read enquirer records and student member records as either of these may become members (students after they have passed the examinations, and enquirers if they are already exempt from the examinations).

The program must also be able to create new member records

(as well as create accounting entries for the subscriptions etc.). The program must also indicate that a particular student is a student no longer. A practical decision has to be made about whether an old student record is to be copied into an archive, and/ or physically deleted from the files, or kept in but merely 'tagged' to show that it is no longer current.

There is a danger in physically deleting records in case something may be wiped out by mistake. Thus many systems will choose to add a 'delete tag' to a record to show that it is not available for use. Only when suitable checks have been made will records be physically deleted—often by a special supervisory program.

There may be an additional security feature within a program or before calling up a program which requires the user to enter a password. Different passwords will confer different entitlements to data access. The *actual* passwords would be closely guarded.

A complete program/entity matrix must be drawn up showing for each program whether it is permitted to Insert, Amend, Read, or Delete a record, or whether it is permitted a partial amend or read. If a program is only permitted to access certain fields of the record then these must be specified.

This matrix is very important for two reasons. The most obvious one is security. The users must not, accidentally or on purpose, have unauthorized access to data. The second reasons is integrity. It is important for the database administrator to be sure which programs access which data so that she/he must know how to deal with any errors that have been detected in the system (e.g. who might have made calculations based on erroneous data). Also if it is necessary to change the structure of a data record during some future enhancement of the system then she/ he must known which programs might be affected.

There are sometimes restrictions on the sequence on which programs may access data. For example you might want to record all the day's subscriptions receipts before printing subscription reminders, just in case the offending subscriptions had in fact been received. Sometimes the sequence is essential, as on a reservation system you could not start to book a seat before having found out that a seat was available. Otherwise there wouldn't actually be a seat there to book.

8.5 CONSTRAINTS ON THE SOFTWARE

As well as the logical requirements there may also be practical requirements. These can be associated with:

(1)   **Size.** How large are the files going to be? How many records of each type? As databases get bigger so the technical problems associated with locating a particular record increase. A large database might well keep an index to the index to cut down search time.

(2)   **Volatility.** How often are old records deleted and new ones added. A volatile database is technically much harder to manage then a stable one and may require more sophisticated software to handle the 'invisible' functions such as 'repacking' the data on a disk after gaps have been left in it by deleted records and new records tacked on the end.

(3)   **Response time.** There are two aspects to this. Occasionally there is an absolute limit to the time one can wait for a response from the computer. For a robot or a computer-controlled factory this may be a fraction of a second. For a telephone enquiry system, the person on the other end may wait a few seconds. A customer standing at the counter and waiting to know if an item is in stock may wait a couple of minutes. In more complex queries a user might be delighted to get a reply by the start of business the next day, although even this could be a constraint on a rarely run program. The other aspect to response is the total time taken to process a large number of transactions. This is a simple figure to compute on a single user system, but where more than one program is running at the same time, there is a traffic jam problem. This is noticeable at busy times even when the computer may appear to be only processing 50–70 per cent of its maximum number of transactions.

There are mathematical formulae available to help determine the particular hardware/software requirements in difficult cases. For the small business system it is usually enough to know the average and worst case figures for each of these. For example the Professional Institute could say that on average there were 55 membership applications in a week, but that there might be 300 in the first week in August when the examination results come out.

8.6 SOFTWARE SELECTION ISSUES

In these days of readily available and often well-known software packages it is very tempting to look at a situation and say 'The XYZ package will do this.' An analyst with experience of that situation might come to such a conclusion and be able to implement an acceptable system on it. This is rather like an experienced surgeon looking at a lump and saying 'That's not cancer'. The person who suffers if he is wrong is the patient, in our case the user.

Thus even where a solution sprang to mind it is as well to finalize the requirements if only to show that the obvious is justified. Usually, however, the choice is not entirely obvious, and there is a choice between different permutations of:

(1) Existing software package.
(2) Existing package tailored to specific requirements.
(3) Program generation (which elicits simple program and data specifications and generates programs to fulfill the specifications).
(4) Bespoke software which will require detailed design work, coding testing and documenting.

8.6.1 Software Packages

The advantages of a software package are obvious—it is readily available, relatively cheap, thoroughly tested and documented, and it is supported by a vendor. It is also possible to see it before you pay for it, which is a very important consideration, as many users feel that they did not get the software they thought they were paying for. Rarely will a package exactly meet your requirements. The question to be answered is whether you could write additional programs to integrate with it to meet the requirement, or whether you could get away with redrawing your system boundaries with more or different clerical functions.

8.6.2 Tailored Packages

If a package seems too rigid then you may go for a tailored package. This is likely to be more expensive in the first place as it is not sold in bulk. There will be an additional cost of so many man hours (or days) to tailor it to your particular requirements. This is becoming increasingly popular as software suppliers are identifying more types of application—accounting, reservations, bill of materials etc. They combine much of the ready availability and provability of the package with a level of flexibility that is adequate to the needs of an increasing proportion of the market.

8.6.3 Application Generators

Where your application does not fit into an existing mould then you may be able to get an acceptable program suite from a program generator. This is a program which accepts specifications of data structures, screen and printer layouts and procedures and then converts them into programs. Quite a lot of R & D is being done on these at the moment. Some of those currently on the market leave a lot to be desired, especially as they tend to generate a rather cumbersome program code. But if you want to set up a straightforward program quickly and cheaply, and do not mind it being a bit slow in running, then you might benefit from one of these generators.

One area where they are finding increasing use is in building prototype systems. This gives the user a chance to see whether the prototype fits his requirements and to make any changes that are required. Once the form of the program is complete it can be handed over to a programmer to rebuild in more efficient and reliable code.

Somewhere between application generators and customer built programs, and overlapping with both of them, lie what are known as 'fourth generation languages' (Forgols for short). The term covers a variety of different methods of specifying an application in a language that is closer to the application than conventional languages such as Cobol and PL/1. This makes programming quicker than in conventional languages, and allows the programmer to concentrate more on the application and less on the technical details of the implementation.

8.6.4 Custom Built Programs

Only if all of these options fail should you consider having programs specially written for you. In principle you can get exactly what you want in this way but in practice the difficulties and delays of detailed specification, coding, testing, and documentation cause dissatisfaction. A great many man-hours are required —from programmers, analysts, users and operators—and a lot of computer time is used up in testing. During the delay the user will have been deprived of the use of the program and often the actual requirements of the user's perception of them will have changed. An alarming proportion of software is thrown out as being unacceptable on delivery, or else requests for changes are submitted almost immediately. Much of this can be avoided by the use of prototypes which permit the external features of the program (i.e. what it appears to do) to be sorted out and then fixed while the programmer concentrates on technical problems of ensuring an efficient and reliable implementation.

8.7 SOFTWARE VENDORS

Software is rather like a car in that it needs to be maintained after it is sold. Perhaps it is actually more like a house in that it can be extended and altered as well as repaired. There are various stages in this process:

> ensuring that the software is correctly installed on the particular machine. Even on a micro there are a variety of disk formats, printers etc. that may be used, which all need to be tied in to the program. On a larger computer there are:

>> many possible permutations of equipment that need to be taken into account,
>> providing good documentation which is complete and also readable,
>> training users so that they understand not only which buttons to press, and what the sequence of operations is, but also how to ensure the security of the data and how to get the best use of the system,
>> informing programmers of such things as file layouts so they are able to build additional programs to extend those available in the package,
>> correcting 'bugs' (i.e. software malfunctions) which may be observed in the system,
>> providing enhancements to the software either at the time or later,

providing new versions of the software for new operating systems. This is particularly important for mini and mainframe systems where suppliers may refuse to support old versions of the operating system,
helping users transfer from their old to new versions of the software, or even from old to new machines.

It is essential that a clear decision be made about how much support is required in a particular situation and how much it is worth paying for it. There is an enormous range of possibilities. At one end there is the person who only wants a micro package that he can install himself and a textbook from some computer-user club on how to use it. (These books are often much more informative than the manual that comes with the software.) At the other end of the scale there may be a large company committing all their essential data to a DBMS, who are going to want support for a decade or more ahead to help them not only over their implementation but also every change in their activities, and every enhancement in their computer or its operating system.

Different vendors will be able to offer different levels of support. There will also be a variety of pricing structures to reflect the favoured activities of those vendors.

Finally there is the problem of the short lifespan of many computers companies. It is necessary to use an uncommon degree of business foresight to determine which companies are even going to be trading in five years time.

8.8 EVALUATING SOFTWARE

If a software package is offered as a solution to the user's problems, or if there is a proposal from a vendor to tailor a package, then this must be evaluated to determine its suitability. In a way this is the software design process put in reverse.

Firstly does the software meet the technical requirements? Can it handle the required functions, manage the volumes of records, and the numbers of transactions? Does it support the right sort of database, with a suitable query language? Does it offer the right levels of security? etc.

Working backwards from this, does it offer the right man-machine interface? Are all the necessary dialogues there, or can they be added? Are the dialogues in the right style for the people who will be running them? Does it support the desired communicating devices, e.g. mouse, or OMR?

Will it fit the socio-technical requirements? Will the operators be able to fit the system into their jobs? Will the equipment it runs on be acceptable from an ergonomic point of view?

Finally, is it actually the software best suited to this type of organization? Was it 'people like us' that the software designers had in mind when they were creating the system?

Very rarely will the answers to any of these questions be a straight 'yes' or 'no'. The evaluation will be a list of which functions it performs, which it does not and which it does differently, which records are built into it and which are not, and so on. This can be done for all the technical requirements.

When it comes to the human factors, the evaluation will be a list of pros and cons. You can describe how work would be fitted around the system, where this would be helpful to the staff doing the work and where you think it might cause problems, also how much training would be required. You can point out where the dialogues are quick and easy for the regular user and where they seem laborious. You can point out features which are supportive of naive users and, where people are likely to use the system without training, what difficulties they might experience.

The question of whether the system was designed to support a particular human activity system can be determined by a judicious study of the advertising material, the manuals, and the vendor. Certain vendors have tended to specialize in certain markets. Products are always geared to a market when they are designed. For example IBM normally deal with larger, well established companies who have their own DP departments, although they are now entering the personal computer market. Wang, Olivetti, and Phillips are primarily concerned with office products. Micros are often aimed at the small business, but often with the assumption that small businesses are rather 'amateur'. This would not be true of, say, a professional partnership which is perhaps where the IBM micro is intended to come in. You must make your own decision.

You can also look at the system that is being offered and the people who are offering it to see whether it 'fits'. Are you comfortable with the whole idea of it?

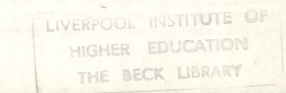

8.9 SUMMARY

The first decision that has to be made concerns how much of the information processing is to be done on the computer. From the functions which were identified in the information model described in Chapter 4, analyst and user must specify those functions which must be done by the computer, those which might usefully be done, and those which are felt to be impractical for reasons of technical difficulty, expense, or unsuitability.

A statement of the logical design of the software (i.e. what it must do) has been developed from the information model, and socio-technical design, and the man-machine interface design.

To the logical design are added some of the practical functions which have to be carried out in order to operate the computer system. These will be worked up in more detail when a particular computer is chosen. At the moment this is a general specification including such functions as 'take back-up copies of files' which have to be done on any computer system but would not appear in a specification of a manual system.

A list must be drawn up of which programs are to access which data items. There are different types of access—read a record, create a new record, update an existing record, logically delete a record, physically delete a record. This list should be extended where programs have access to only some of the fields within a record. Where some programs or records are to have password protection this should be stated.

The choice for software is between an existing package, a package which can be adapted ('tailored') to your requirements, a package produced by a program generator, and a suite of specially written programs. (Advanced programming systems called fourth generation languages may cover either of the latter options or lie roughly between them.) Each of these possibilities must be considered in general terms to see if it is worth pursuing further.

The evaluation of a software package consists of taking it backwards through the stages in the design process and listing out the extent to which it meets the functional requirements, supports the required data structures etc. Then there must be a discussion of the way it fits into the working life of the organization, the training requirements, the suitability of the dialogues for the people who will be using them, and the way in which the whole system fits into the organization whose needs it is to serve.

At this point it is possible to take a general decision about whether or not to go for a software package (including the extent to which it would need to be tailored'. If there is nothing suitable on the market then more detailed design work must be undertaken.

9
Opportunities and Constraints of the Computer

Computer-based information systems (CBIS) are fundamentally different from manual systems. They are more rigid in their procedures but far more flexible in the number of different procedures they can perform on the same data. In many ways they are far more accurate than human clerks but when something does go wrong, it is often much more difficult to put right. Once a decision has been made to implement a CBIS, then both the opportunities and constraints need to be explored.

9.1 ADDITIONAL INFORMATION RETRIEVAL

Once information is stored in a computer then it is in principle much easier to extract selected bits of that information than it would have been if the information had been in a filing cabinet. Various possibilities are:

(1) Information can be sorted and presented in different sequences (e.g. library books can be printed on author, title, or subject sequence).
(2) Records which meet certain criteria can be selected out for presentation (e.g. Print all the books by Smith A. W.).
(3) Statistical calculations can be performed (e.g. average sales figures over different periods, trends in sales etc.).
(4) Cross references can be made between different types of records (e.g. ice-cream sales figures and meteorological reports.).

While these can all be achieved in principle, they will not always be easy to achieve in practice. Users are often bewildered by the ease with which they can get answers to what they had supposed to be a difficult problem, and yet on another day an apparently simple request will prove impossible. This is because the way in which the data has been structured and stored determines the way in which parts of it can be retrieved.

For example the hierarchical record structure for sales information shown in fig. 7.2, makes it very easy for the customer records for one salesman to be cross-referenced, but to produce statistics for all customers would involve going region by region, and salesman by saleman, to retrieve their customer records before you can do anything with them. This arrangement is ideal for helping a salesman keep track of his personal customers and also for comparing one salesman's performance with another's. It is not so suitable for working out company-wide sales trends.

This means that thought has to be given to potential requests for information. This consideration comes after the essential elements of the system have been defined, and after a decision has been taken to design the software rather than buying in an existing package. At this stage we are considering the possibilities and the difficulties for users that will be thrown up as we go through the process of designing a specific information system.

9.2 WHAT INFORMATION IS TO BE RETRIEVED?

A basic specification has already been agreed with the users. What we are asking at this stage is 'Given the data that is going to be in the computer, what additional information retrieval facilities would be desired'. At this stage we are looking at information obtained by manipulating existing data. Users often have to be helped to appreciate the possibility of getting hard information when they had always been obliged to use guess-work. However, the more enthusiastic users must realize that they cannot get all the answers at the touch of a button. They must make realistic appraisals of what additional information they would like to receive regularly and what they might need on occasions. Values and priorities must be assigned to each of these. Urgency is also a factor. Some information is well worth waiting for, but there are occasions when a decision has to be made by a deadline. In this case the information would only be useful if you can get at it in time to do something with it before the deadline.

If the analyst is aware of these additional demands before the main database design has been completed, then some or all of these additional requirements can be designed into the system. Much valuable information may be obtained in this way for little marginal cost.

9.3 SUPPORTING DECISION MAKING

In Chapter 1, a decision support system (DSS) was described as being inherently semi-structured. This is because one of the truest tests of the decision maker is knowing which factors to take into account. A true 'expert system' would be able to deduce all the factors. However, we are not asking the computer system to make the decisions. What we are saying is that where the raw data is available in the computer system then the decision taken should be able to retrieve such parts of this as will help.

Only where decision making is both complex and crucial is one likely to create a CBIS primarily for decision support. Normally the DSS works on data that would be available as a by-product of a routine information processing activity. However, there is also the possibility of putting some extra data into the system that can be used for more informed decision making. An example of this on a normal form is the box on the bottom of many a job application asking 'where did you find out about this post?' This is to help control expense on advertising.

In the Video Shop it might have been useful to analyse viewers' preferences in order to decide on a better stocking policy. A simple way to do this would be to include a single letter category to the video record. This would have a negligible effect on file size or time taken for data entry. A few extra statistical records would have to be kept within running totals of the numbers in each category tried out. Finally, a few extra lines of code would have to be added to the procedures for hiring a video to update the statistics.

The cost of implementing this DSS at the start would be negligible since it could have been incorporated into the original program. However, it would have been a substantial exercise to add it to a system which was already up and running.

The important task is to define the statistics which would really help make the right stock choices. This is the *real* decision, i.e. 'What factors are important in predicting demand?'

9.4 A WORD ABOUT STATISTICS

Statistics are always treated as facts, but they are not. They are interpretations of the significance of data. For example if the Video Shop statistics showed that the most popular videos one month were sentimental movies, that is a fact. However, it doesn't *mean* very much. Maybe there are more sentimental movies in stock than anything else, maybe the month was December and everyone likes watching sentimental movies over Christmas, maybe there was one such movie that was getting a lot of media coverage at that time, maybe they had just done a big promotion at the 'WI'. Thus there are still a lot of questions about other factors that ought to be taken into account even when looking at those figures.

But that isn't the end of it. You need to get a feel for how things have gone over a specified time. For example is there normally a demand for sentimental movies at Christmas and if so is it worth getting in some extra ones to meet the demand. This is an example of 'seasonal factors'—toys at Christmas, ice-cream in the summer, raincoats in the autumn, cars in August when the new registration plates come in. The best thinking in statistics indicates that you start with a theory about how probable a certain event is. For instance there is a 70 per cent chance that we will sell at least X tons of ice-cream in August. This is based on your experience and on whatever data you have managed to get hold of. You base your decision-making on this figure in the first place and then adapt it in the light of further experience. If you sold more than X tons, the probability of doing so another year in the future might be 80 per cent and you might consider making more.

Returning to the Video Shop, the whole DSS was based on assigning a category code to each video as it is entered into the system, but who is to say that these categories represent the most meaningful classification. Perhaps the choice has more to do with the leading actors. Also, who is to say that a film has been correctly assigned to its category? There will be many films which have aspects from more than one category, such as comedy and sex, or which tend to defy description because they are 'in a class of their own!'

You can probably see why the Video Shop owner decided to stick with his own judgement. However, as the number of records increases, one's ability to make reliable judgements decreases. Statistical theories related to decision-making have been progressing at least as fast as the abilities of computers to calculate those statistics. This means that many users who have been using 'statistics' for a long time may find that it is possible to extract much more meaning from them by taking advice from a statistician.

This must always be done before the data structures and statistics collecting procedures have been implemented. Otherwise much of the 'meaning' of the data may have been lost by omission or distraction.

9.5 COMPUTER SECURITY

This has been mentioned in respect of database systems.

If a computer system fails for any reason, then the users could find themselves unable to carry on their work. This could be a complete disaster for the company especially if it took too long to get the system working again or to recover lost data. Even a minor failure could lead to hours of work being lost. There are various aspects of computer security, such as:

precautions in case of hardware failure,
detection and correction of errors in the database,
guarding against careless mistakes by operators,
prevention of unauthorized access,
contingency plans in case of disasters like fire.

In some ways this is the responsibility of the computer manager, but the analyst has a duty to ensure that the user has considered in advance the security aspects of the operation. These include:

What data is confidential and how much is it worth in terms of additional security measures to keep it so?
Which functions are absolutely vital to daily operations and which could be omitted or delayed if there was a shortage of computer power?
How long a delay could be tolerated before the computer was

fully functional and how much would it be worth to the company to reduce this time?

What arrangements should be made to keep things ticking over if the computer were down?

Where will the computer be situated and are there any risks associated with that location (e.g. floods in basements).

Is all the computer equipment secure against theft, forcible entry etc.

What about fire precautions, insurance etc.

One area of growing concern is computer fraud. Breaking into computers is the latest criminal activity. Teenagers doing it as a test of their intellectual prowess may do damage, while a sophisticated criminal could steal company secrets or cream off money into his own bank account. A disgruntled employee could even sabotage the database or the software library.

9.6 SECURITY IN OFFICE-BASED SYSTEMS

Large computer departments have been learning a lot over the last few years about how to protect software and databases from computer failures and other disasters. Many users would feel that computer departments still have a lot to learn because there are few systems as reliable as their users think they ought to be. This may be because no one had given enough thought or paid enough money to create a really robust system.

Now that computers have moved out into the user offices, a great deal more thinking has to be done about security. What is more the users will only have themselves to blame if this thinking is not done in time to protect their systems from failures. Here are some questions that ought to be asked:

Do you take back-up copies of all major WP documents or microcomputer databases?

Do you have a procedure for recording when these copies were taken and where the backups are stored?

If your originals were destroyed would you know how to bring the copies back up to date with all the latest amendments?

These security procedures are normally automated on mini and mainframe systems, but in the office they actually rely on the typist taking the copies etc. This imposes a discipline on people that can seem quite irksome at first. Designing these procedures can be difficult for the users if they have no first-hand experience of what can go wrong, so they will need help from the analyst. Users need to be trained in these procedures as part of training in the use of the new system. When it comes to maintaining security, they are on their own and must enforce the taking of backup copies etc. until it becomes automatic—just as a good typist automatically takes a carbon copy of whatever she types.

The following procedures are useful for WP security:

Keep a log of all major documents showing who typed them, when they were last edited, what the file name is, who it was for.

Set aside a time each week (at least) to make sure that all the backup copies are up to date.

Keep back-up copies in the safe (preferably a fireproof one).

Maintain a good filing system for the live disks.

If you have a microcomputer system with a hard disk, then taking back-up copies is a problem because it can take 20 minutes to dump the whole of a disk to floppies. This problem has not yet been solved by the manufacturers although some offer a system in which there is one fixed hard disk and one which is removable to hold the copies. You just have to decide what you would do if the disk drive did fail and you lost all your data. Just remember that things are likely to go wrong when you are busiest because that is when people cut corners and get careless. You really have to decide how much that security is worth.

9.7 CHANGING THE STYLE OF WORK

Once the decision to install a certain type of computer system has been taken, then a lot more thought can be given to the way in which people are going to arrange their work. This will have been decided in general terms during the socio-technical design stage, so users will know whether they are going to be inputting work at a terminal, collecting work into batches, or whatever. Now they will need to think about a whole variety of details both in the operation of the work (i.e. exactly how each task will be done) and also in the physical arrangement of the equipment in the office.

Many of the aspects of the operations will be the concern of the users themselves. They will have been involved in the design of the dialogues. As the detail of the design is fleshed out so they must continue to be closely involved. This is another reason why a prototype system is so valuable. Users can try out what it will be like to use the system in practice. This will give them the best possible start on the task of planning the new regime.

This planning has to be done in several areas:

the organization of the work,
assigning particular tasks to particular individuals,
creating a training strategy,
determining and then setting in motion any structural alterations to premises, including electrical and data wiring,
designing a comfortable working layout for terminals, desks, printers etc.,
ensuring suitable lighting, ventilation and acoustic insulation (especially for printers),
designing a comfortable and efficient layout for the work-stations themselves.

This work starts before any actual hardware orders are placed or the software details finalized. It carries on throughout the implementation phase as the system is delivered and integrated into the work of the office.

9.8 USER TRAINING

The dialogues necessary to achieve each task were designed in the man-machine interface discussed in Chapter 6. A suitable method of implementing the dialogue was also selected, for example menu, on-screen form filling etc. The detailed method of carrying on that dialogue will be determined by the selection of a software package or by the writing of the software. Users should always be involved in the detailed specification of the parts of the dialogue which they will be using. Plans need to be drawn up for user training before, or at least alongside, the plans for implementation of the system.

It is quite clear that all potential systems users must be identified and trained in the use of those dialogues which they will actually be using. The difficulty lies in the timing of this training so that they know what they are supposed to do before they are called upon to do it. However, if training starts too early, users will not have had the opportunity to consolidate what they have learned by practising it, so they will forget some or all of it.

To be really effective, training needs to cover more than just the manner in which individuals must perform their own specific tasks. Users need to:

- understand the equipment; screen, keyboard, printer etc.,
- know how to operate the actual dialogues and procedures,
- have a mental map of what is going on inside the system,
- see how the system fits into their work.

Different individuals and certainly different types of user will learn these in a different order. The analysts and programmers will certainly be working from their own mental map. Managers and decision-makers will probably want to start there too. Users who will be running routine tasks on the system will probably feel most comfortable if they start with an introduction to the equipment and then the dialogues. However, this is a generalization, and no-one will feel entirely comfortable until they have developed an understanding of all four aspects.

It is impossible for instructors to cover everything in one training session, and even if they could, the trainees wouldn't be able to take it all in. Many things that are covered in training courses do not 'sink in' until users come across them in practice. An extensive European study of user training showed that only 50 per cent of users were satisfied with the training at the end of the course (the average duration of the course was $\frac{1}{2}$ to 3 days). However, this figure for satisfied trainees dropped to only 30 per cent after a month as users battled with their ignorance of the new system.

Infrequent users of the system pose a particular problem in that training has to be provided as and when it is required.

Various training strategies may be used. For instance:

Introductory sessions plus detailed training for each task as it is computerized.
A cadre system where one user from each department is very thoroughly trained and then sent back to instruct colleagues and be there to sort out their problems as they arise.
Good user manuals and a thorough training in what they mean and how to get information from them.

Visiting trainers who demonstrate the system at the workplace and help users actually set it up and working.

Access to trainers after installation to solve problems that do not seem to be covered by the manual.

A 'drop-in' training facility for irregular users who want introduction to parts of the system they have not used before.

10
Hardware Selection and System Implementation

At one time these used to be the central concerns for the analyst. The search was for a machine that could perform a certain task and for ways of getting it working. Now the emphasis is not so much on the machine itself as on the whole range of associated technical facilities. There are various factors to this:

The machine must be large enough and fast enough to process the work.

The possibility of upgrading the processing power or internal memory of the machine if the volume of work increases.

The range of peripheral devices (printers, disk drives, terminals, graphics work-stations etc.) that can be attached.

The operating system offered and whether this will support the required software.

The type of networking offered, both local area and wide area, and the particular standards used.

The support offered by the vendor in the short and long term.

The possibility of transferring present software and data files to the new machine.

The compatibility of the proposed system with other systems in use in the organization.

After the socio-technical design phase a general decision was made about the type of computer system to go for. This general decision must be supplemented with additional information before an invitation to tender can be prepared.

10.1 THE REQUIRED PROCESSING POWER

The function hierarchy from the information model discussed in Chapter 4 was expanded in Chapter 8 to provide a software specification. This included not only the processes needed to perform the user's work, but also the 'housekeeping' processes needed to support the computer system. The dialogues described in Chapter 6 give the number of transactions between user and machine to perform each task. An estimate must also be made of the amount of processing involved in the tasks performed within the computer, such as sorting files into a different sequence to prepare a report, performing statistical calculations etc. Put together, these tell us WHAT processing has to be done.

We also need to know HOW MUCH processing is to be done. How many transactions will be processed on an average day, and on a busy day? How many transactions will be processed during the busiest hour? How many reports are likely to be produced on a busy day and how many statistical calculations? Are the busy times for all these different jobs likely to come at the same time?

The number of transactions has to be multiplied by the time taken to process a transaction in order to determine the power required. This will give a working estimate which is usually enough to choose between the processing options offered. It may need to be worked out in more detail by technical specialists if there is any doubt about the ability of the machine to cope with the work.

There is no point in expecting the machine to work at more than about 70 per cent of capacity even at the busiest times, because any increase above that will result in frequent 'traffic jams'. The computer doesn't take any longer to do the actual processing, but an individual transaction will have to 'queue' after it has been entered at the terminal to get into the processor. Once processing has started it will have to queue to get access to the disk for the required information and then for a chance to get back into the processor again. This will be repeated several times during the course of one dialogue and the accumulated delay time may well become noticeable to the operators and may degrade throughput quite considerably.

Processing power is relatively cheap these days not only compared with what it was, but to the other costs in the computer system, notably manpower. Also users have been accustomed to expecting, even if not actually getting, prompt responses at the terminal. Thus it is often wise to err on the side of caution when deciding on the required machine power unless this would seem to indicate a quantum jump to a much more expensive machine. If this is the choice posed then more careful calculations of requirements will have to be made. The advice to buy a larger processor than seems absolutely necessary is often not followed because the cost of the computer is more visible than the larger costs of people waiting around for computer responses.

If a computer system is successful, then users will certainly think of additional tasks for it to do. This means that there will be calls for an expanded service. If this is put on the existing machine it could slow things down. The necessity for expansion may well be anticipated in the original specification, if it is the first stage of some larger plan.

Expansion can be achieved by:

Adding more memory.
Adding one or more additional processors to a network.
Adding high speed disk and using a 'virtual memory' operating system which parks unused bits of processing temporarily on the disk, thus allowing more space for whatever happens to be active at that microsecond. Work can be rolled on and off the disk with only a fractional increase in total processing time.
Using intelligent rather than dumb terminals to shift some of the processing of the data entry and display to the terminals and away from the overworked central processor.
Swapping to a larger machine in the same family. This is only possible if the original system was 'upwards compatible' with the larger one. Even so it is an option of last resort as it will involve considerable expense and disruption even if it does not require the programs to be rewritten or the database re-created.

10.2 PERIPHERAL DEVICES

The most obvious of these are the terminals. The questions that must be asked are:

How many terminals are going to be needed and what facilities should they have?
Should the screens be high, low, or medium resolution; will they need graphics capacity, colour, light pens, touch-sensitive screens?
Are these for occasional use or must they match up to the

standards of health and safety for continuous use, e.g. clarity, lack of glare, tilt and swivel, detachable keypads etc.
Should they be intelligent terminals?
Should they have their own disk drives, printers etc?

Disk drives are themselves peripherals in a sense. Many micro-computers have the disk drives built in, although there is often the capacity to add an additional drive or a larger hard disk. On larger computers there will be considerable choice in the speed and volume of disk storage attached to a given computer.

The volume of disk required is determined by multiplying the number of records of each type by the number of characters in each record. This number of characters ('bytes') should then probably be doubled to allow for odd gaps between records, and for the space taken up by indexes and pointers from one record to another associated record, for example from 'customer' to 'rental' in the video shop. These calculations can be done much more exactly where necessary, using information provided by the manufacturer about the way in which records are packed into the space available. The space can also be optimized if records are fitted neatly into the existing disk organization, e.g. many micros split disk space into units of 128 characters, so that ideally records should be just less than this.

Printers can be grouped into:

High speed 'line' printers which print a complete line at a time. These are expensive, fast, noisy, and the results are not very neat.
Dot Matrix printers which are cheap, quiet, reliable, moderately fast, and capable of printing different characters and graphics. Print is usually more or less 'dotty' depending on how many dots are printed per character.
High-quality daisy-wheel printers which are slow and noisy but which produce 'letter-quality' print, and even 'camera-ready copy' which can go straight to a print-shop for photo-lithography.

The printer is probably the slowest part of the computer system and is likely to prove a bottleneck for any system where much printing is done. There is a lot of talk these days about the paperless office, but if anything we are producing more paper these days rather than less. Therefore it is important to work out how many pages of different types of printing are going to be produced in order to work out how many printers should be built into the system.

There are various types of graphics plotter available.

Input devices which may be considered are optical character readers, optical mark readers (OMR), wands for reading magnetic labels and bar-codes. The prices of OMRs and bar code readers are coming down to a point where they are widely available and suitable for many routine operations.

10.3 OPERATING SYSTEM AND NETWORKING

Until recently each computer or at least each family of computers had its own operating system. This changed most obviously with CP/M, which was for a whole range of microcomputers based on the Z80 microprocessor. With minicomputers the same thing happened with UNIX, which has been deliberately constructed so that it behaves similarly towards the users and the software whatever machine it is working on. This is a popular move and is already being followed up.

There are, of course, all sorts of technical questions to be asked about the hardware. However, as far as the user is concerned, there are two major questions:

- Does it support the required applications software or the intended software development tools?
- Does it support a suitable telecommunications and computer network system?

10.4 MANUFACTURER/SUPPLIER SUPPORT

Large computers may be purchased directly from the manufacturers, who have regional sales offices for the purpose. Microcomputers are likely to be purchased from a supplier who may have got them from a wholesaler, so the user could be two removes from the manufacturer.

For DP departments dealing with computer manufacturers the negotiations are hardly straightforward, but they are fairly well established. In the case of dispute it is possible to refer the matter to an independent expert for arbitration. Matters to be agreed include:

delivery dates,
acceptance tests to show that the equipment is fully functional according to specification,
maintenance agreements and the availability of spares.

It is a standing joke (or a permanent scandal) that microcomputer manufacturers sometimes seem to give the impression that they design their advertising before they design their micros. Availability of equipment and especially of their enhancements (such as hard disks, telecommunications etc.) is notoriously difficult to guarantee. This is due to optimistic marketing, high levels of demand, and the shortage of components and manufacturing expertise.

Where delivery time matters it is worth ignoring discount prices and going to more upmarket machines and suppliers although this is no guarantee unless the whole of your required system is literally on the shelf. There is a great temptation for buyers to go for the very latest machine because its technical specification is so amazingly good. However, unless you are certain that all the software and enhancements are available with the new machine, then perhaps you had better settle for a well-established machine.

Maintenance contracts for computer equipment may cost 10–15 per cent of the cost of the hardware. This would cover the call-out of an engineer and the replacement of faulty components. The price can be varied according to the speed of call-out. Some critical systems may justify a 2-hour call-out, others may put up with 24 hours. With microcomputers you can save quite a lot of maintenance money by bringing the machine back to the shop for repair. This may be acceptable provided the repairs can be done promptly. The availability of maintenance may be a critical factor for users outside the major towns.

Support continues throughout the life of the system, including the addition of updates and enhancements and the transfer to a new machine when the old is replaced.

10.5 SYSTEMS IMPLEMENTATION

More books have been written about this than almost any other computing topic. For the microcomputer user implementation means little more than taking the machine out of its box, plugging it in, loading the software package and then typing in the existing data. Even so, you may be advised to have this done by the supplier simply to shortcut any arguments about whether the machine did or did not work when supplied. For larger systems the installation of the machine must be done by the supplier or his agents. The whole implementation process will involve a complex piece of project management which is worth a book of its own.

The emphasis on systems implementation has been reduced in part because we now know so much more about the work that has to go into the design phase. This means that the programmers etc. can have a much fuller description of what is required. They also have much more expertise and better software development tools. Much of the technical side of programming is covered by packaged software—operating systems, database management systems, tele-communications systems, fourth generation languages etc.

The task of the analyst is now to provide acceptance criteria to demonstrate that a particular system meets the user's requirements. The requirements have been spelled out in the design of the human activity system, socio-technical system, and man-machine interface. The details of the testing will depend on a simulation of user activity of increasing difficulty:

simple transaction,
normal activity,
high volume activity,
activity full of errors and exceptions.

These tests may be based on a sample of real-life activity, e.g. some portion of a previous day's work. Methods of checking the computer's results for accuracy must be chosen. This is where running the computer system a few days behind the previous system is very helpful, although it may not be easy to manage especially as it means processing a lot of the work twice.

10.6 INSTALLATION LIFE CYCLE

Not so long ago the majority of computer systems being installed were to replace manual systems. For large computers or large companies this is almost certainly not true any more. Only where micros are going into small businesses or offices remote from a data processing department is this likely to be the case now. There seem to be three phases that an installation goes through:

1. Experimentation where the first computer system is brought in to perform a specific task, and the organization learns (often by its mistakes) what computer systems can do.
2. Consolidation where the organization's routine activities are gradually transferred to computer and the computer system is upgraded and expanded to cope.
3. Maturity where users take an informed interest in computerization, start experimenting with information retrieval for their own purposes and requesting the construction of decision support systems.

Thus the implementation of the new system will often consist of an extension to an existing system, or of the building of additional information retrieval and other decision support systems on top of an existing system. Care must obviously be taken not to corrupt the old system during testing of the new one, or to disrupt it in any way, even by slowing down response times to an unacceptable level.

10.7 SUMMARY

Hardware selection is not so much the selection of the right computer as of the supplier offering the right range of technical facilities based around a suitable computer. The computer must be of the right speed and power to perform the type and amount of processing required both now and in the foreseeable future (possibly after an upgrade), but it must also:

support the required software including languages and data-bases management systems,
support the networking that might be required now or in the future,
fit in with existing systems and offer a means of converting existing programs and data files,
be support for the required level of maintenance,
support the required numbers and types of peripheral devices and the volume of disk storage.

The implementation of a system may consist of merely plugging in a micro and typing in existing data. For a larger system it may involve a programme of alterations to buildings, re-wiring or installation of data lines to terminals, installation and testing of the machine, program coding and testing, and finally systems testing. The analyst may be involved in program testing by providing sample data and checking the output. He or she will certainly be involved in systems testing, making sure that the system being implemented meets the user's requirements. In many cases the system being installed will be an addition to an existing system such as additional query facilities on an existing database, or an extension of a system to include a new area of work.

11
Overview of Systems Analysis

Some of you may prefer to read this chapter first. People divide roughly into those who think 'top down' and those who think 'bottom up'. Top down means that you like to get the broad picture first and then work down to the details. Top down thinkers should read the overview first. Bottom up signifies a concern with understanding the details and then seeing how they fit into a broader picture. Both these styles of working will be encountered by the systems analyst. Indeed there are great arguments about which approach the analyst should take himself! In a way it is his job to reconcile the two; to draw up the broad picture of the information system and the way it fits into the organization, and also to fit together the details both for the users and for the programmers.

11.1 SYSTEMS ANALYSIS THEORY

With the very earliest computers, only the hardware was thought to be of any theoretical interest. Programming was thought to be a clerical detail, which was how the first programmers came to be women. As computers advanced so did theoretical study of programming. Many techniques, such as structured programming, emerged as a result of this theoretical work and is now generating a whole new discipline of software engineering.

Computer scientists have been heard to defend the view even now that systems analysis is not a discipline but that it is just common sense. These people show that they have never tried to do a serious piece of systems analysis if they do not realize that it is a demanding intellectual undertaking. They are also of the school that doesn't recognize anything as a 'theory' unless it is based on a mathematical formula. But do theories have to be like that?

The first theory of systems analysis was 'The analyst asks the users what they want and then designs a system to meet those requirements, so he needs good technical design skills.' Many 'experiments' were carried out to test this theory, or to put it another way, many systems were designed on that assumption. The ranks of disappointed users showed that this theory was incomplete.

Various ideas were put forward about what else was needed, such as:

Users could not always state their requirements clearly, therefore the analysts had to know about interviewing techniques etc.

Different users had different views of the same data, so data analysis techniques were needed to provide a neutral definition of the data.

Sociologists could provide an insight into the labour-relations problems being caused by the installation of computer systems.

Cognitive psychologists were called in to help when users had difficulty understanding how to get information out of the computer.

As machines themselves became more complex so did the technical skills required in systems design.

As more sophisticated programming languages and database management systems were developed, so some system development tasks became easier.

With the advent of packaged software some adverts proposed that the role of the systems designer was about to disappear.

Many different computer manufacturers, software houses, and large user departments worked out their own approaches to systems analysis and design (SAD). These sometimes started as a philosophy of what ought to be important in design, for example Scandinavian methods tended to start with the needs of individual users. Some SAD methods arose as a response to particular problems, such as data analysis being developed by a company installing large databases.

Teachers watching and often taking part in the development of SAD developed a variety of theories to improve on the original one that systems analysis was all common sense.

11.2 THE MULTIVIEW METHODOLOGY

'Methodology' is a popular term in SAD. It is widely misused because it has a nice scientific ring to it and can give apparent authority to what should properly be called methods. What it means is a theory about a practical discipline (in this case systems analysis and design) which helps people engaged in that discipline to select the best methods of working in that particular situation.

The most important thing to realize about practical disciplines is that you cannot slavishly follow a text-book. Nobody can guarantee 'perfect results every time' by following a set of techniques, any more than having a good cook book means that your family will always like the meals you cook. What we offer as text-book writers is a general understanding of the factors at work in information systems and some techniques for dealing with them. What follows represents the best theory available to date of what those factors are and how they relate to each other.

The name Multiview derives from the different views which can be taken of an information system. It can be seen as part of a human activity system furthering the business of a company, the functioning of a department, or whatever. It is part of a system in which staff work with technology to perform certain tasks. At another level it is a situation in which an individual enters into a dialogue with a computer. While it is all these things it is also based on an understanding of the flow of information around the organization and between the organization and the outside world, the functions to be performed on that data and the requirements for storing information. Finally it is a complex technical system with requirements for hardware, software, databases, security measures, and so on.

The theory that we are proposing is that whatever advances take place in computer systems or in analysis and design techniques, there will still be a five fold view of an information system:

- human activity system
- information model
- socio-technical design
- man-machine interface
- technical sub-system.

and that the framework for inter-relating these will remain useful even when some of the techniques described in this book have become obsolete. The rest of this chapter concentrates on that framework and on the way in which it can be applied in particular problem situations.

11.3 PROBLEMS AND PROBLEM SOLVERS

A systems analyst is called in to solve a problem related to information processing. In some cases, particularly for microcomputer systems, the analyst may also be the problem owner, i.e. where someone is trying to solve their own information processing problems. Normally the problem solver works with the problem owner in the problem situation.

One thing that we all know is that problem situations are all unique. Our theoretical basis relies on the recognition of certain basic patterns which are observed in all situations where people are installing information systems. A textbook can represent an 'ideal methodology' but it cannot show you how to use it in a particular situation. The common-sense theory of systems analysis had some merit. You will still have to rely on common sense to relate the real-world problem to the ideal methodology in order to come up with practical working methods for tackling a particular problem.

The area of systems analysis where common sense, or at least experience, is still all you have to guide you is in closing the gap between the ideal methodology described in the text book, and the real-world problem. There is an inter-relationship indicated in fig. 11.1 between the analyst, the problem, and the problem solving approach. The plan for tackling a particular situation is developed from all three.

Here are some examples of situations where the authors have had to depart from their own 'ideals' in order to tackle a particular problem:

A bid for a government-funded contract where the rules were that any hardware required would have to be included on the bid for the analysis and design work. Normally one would specify the hardware only after all the other requirements had been worked out, but it was made clear that we would have to buy whatever had been named on the bid, or at least something very like it. In particular if we asked for micros we could not later say that we needed a mini. In this case we did a very brief

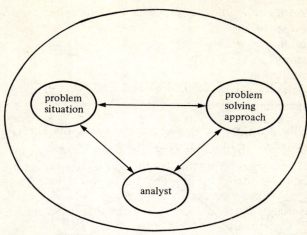

Figure 11-1 Problem solving

outline study of the first four parts of the system (HAS, Information Model, Socio-technical, Man-machine interface) to show that the work consisted of several discrete tasks, each of which could be handled by a micro, and then took an informed guess at a suitable make, disk size, etc. to get the right price.

A records system for a unit that was about to be set up to do work different from anything that presently existed. In this case the analysis of the human activity system was actually a study of what people expected the work would be like, and we started from the boundaries of the system—who they would be dealing with, what activities that would impose, and tried to see what would have to happen inside the unit to support that. The information model was particularly useful here because it was used like an architect's model for a proposed building. The users could state what they thought was required, and then actually try out the model to see whether it could be used for all the information processing and query handling they could imagine being needed.

11.4 THE SYSTEMS LIFE CYCLE

It used to be thought that there were seven stages to a computer system as shown in fig. 11.2:

- analysis
- design
- building
- testing
- installation
- operation
- maintainance.

On to the front of these was put the feasibility study. Various things have caused us to modify this picture.

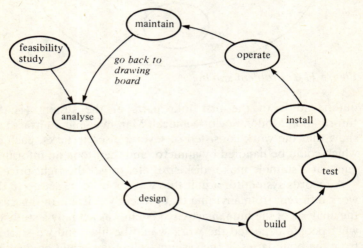

Figure 11-2 Traditional view of the systems life cycle

First, no situation is static, if only because people's understanding of their requirements develops as the work proceeds. This means that you cannot do a complete, once-for-all analysis of the requirements and then carry on with the installation knowing that everyone will be happy. There was much talk of having to 'freeze' the requirements specification in order to let the programmers get on with their work. This doesn't get away from the basic problem that systems design is an iterative process, i.e. you try out one version and then improve it when you see what its limitations are.

Second, the availability of software packages means that you are often not called upon to design and build systems at all, but rather to select them. This calls for a different strategy.

Third, it was difficult to see where the feasibility study fitted in, because it did not seem to be different in kind from the analysis and design activities. Obviously there was no need to build a system and so on, but it was only a matter of convenience to the people authorizing expenditure where the feasibility study ended and the analysis began. Some people would want more detailed feasibility studies than others.

The revised view of the systems life cycle is shown in fig. 11.3. The project may start with a feasibility study or an initial study, either way it sets out in outline what the requirements are. This is followed by a detailed analysis and design of the logical requirements. This covers the analysis of human activity system, information model, socio-technical system, and man-machine interface.

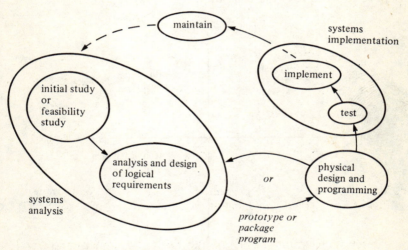

Figure 11-3 New view of the systems life cycle

There are two ways forward from here. You can try out available packages to see whether a system is already available which meets your requirements. Alternatively you can create a prototype system, using an application generator or fourth generation language. This can then be tried out and adapted until it meets the requirements. If the prototype is not sufficiently powerful or robust, then it can still serve as a specification of systems requirements. It is much better for this purpose than a specification on paper. The building, testing etc will then go ahead as on the original view of the life cycle.

11.5 THE MULTIVIEW FRAMEWORK

Figure 11.4 shows a graph with different viewpoints of the system along the two axes. The vertical axis shows a gradation between the organization on the one hand and the computer on the other. Many users have complained that the computer system takes too much account of the computer and too little of the organization whose needs it is supposed to serve. The horizontal axis goes from the people using the system to the technical (or functional) requirements of the system. This reflects what is often the argument of the trade unions when new technology is introduced, that all the concern is with the functions that have to be computerized than with the needs of the people who will actually have to work with the computer.

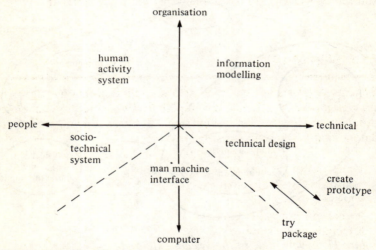

Figure 11-4 The five views of systems analysis

The five parts of Multiview are mapped on to the graph. The human activity system is concerned with the organization and the people. The information modelling relates the organizational and technical requirements. Socio-technical design relates the people and the computer. The man-machine interface is concerned with fitting people and technical requirements to the computer. Technical design is a matter of putting the technical system on the computer.

It is a bit misleading that the 'technical' in 'socio-technical' is used in a very broad sense to embrace both the technology (i.e. the computer) and the functional (i.e. technical) requirements. The people on the 'socio-' side are not terribly concerned about whether the contraints placed upon their work are the result of a limitation in the technology or a limitation imposed by the technical system. As an example one user once asked us why he could not know immediately what the latest figures were for his business. Should he invest in a more powerful computer that could run data entry, update and enquiry programs at the same time? He thought that the constraint came from the computer. It took some time to convince him that he couldn't have any figures until the day's entries had been put through the system. This was because half updated figures are worse than useless, since you may have posted all the credits and none of the debits. The user still felt aggrieved and it still made his job more difficult.

11.6 SUMMARY

Systems Analysis has emerged as a discipline in its own right. A theoretical basis is being developed to give practitioners more insight into the problems they face. The best understanding to date is that systems analysis has five facets:

- analysis of the human activity system
- information modelling
- socio-technical design
- man-machine interface
- technical sub-system.

These are related to each other via a two dimensional view of the organization which has one dimension going from the organization to the computer and one from the people to the technical or functional requirements.

The life-cycle of a system starts with the first four facets taken to an appropriate level of detail for an initial study, a feasibility study, or a detailed analysis and design of the logical requirements. After a full study has been done, either a prototype can be built or an appropriate package can be selected. If the prototype is not sufficiently powerful or robust then a complete system can be built using the prototype as a specification.

In any particular situation the analyst (or problem solver) is working with the user (or problem owner). The analyst must devise a plan of work which matches the ideal methodology into the real-world situation.

Index